Othniel Charles Marsh

The Dinosaurs of North America

Othniel Charles Marsh

The Dinosaurs of North America

ISBN/EAN: 9783743337046

Manufactured in Europe, USA, Canada, Australia, Japa

Cover: Foto ©berggeist007 / pixelio.de

Manufactured and distributed by brebook publishing software
(www.brebook.com)

Othniel Charles Marsh

The Dinosaurs of North America

THE DINOSAURS OF NORTH AMERICA.

BY

OTHNIEL CHARLES MARSH.

CONTENTS.

ILLUSTRATIONS.

THE DINOSAURS OF NORTH AMERICA.

By O. C. Marsh.

INTRODUCTION.

Among the many extinct animals that lived in this country in past ages, none were more remarkable than the dinosaurian reptiles which were so abundant during Mesozoic time. This group was then represented by many and various forms, including among them the largest land animals known, and some, also, very diminutive. In shape and structure, moreover, they showed great variety, and in many other respects they were among the most wonderful creatures yet discovered.

The true place of these reptiles in the animal kingdom has been a matter of much discussion among anatomists, but the best authorities now regard them as constituting a distinct subclass of the Reptilia. Some of the large, earlier forms are apparently related to the Crocodilia, while some of the later, small, specialized ones have various points of resemblance to birds. These diversified characters make it difficult to classify the dinosaurs among themselves, and have led some writers to assert that these reptiles do not form a natural group, but belong to divisions remotely connected and not derived from a common ancestry.

It is not within the province of the present article to discuss in detail the classification of this group, nor to treat fully the various questions relating to the genealogy of dinosaurs, about which little is really known. It may, however, be stated in few words that three great divisions of the Dinosauria are now generally recognized, which may be properly regarded as distinct orders. For these groups the writer has proposed the names *Theropoda*, for the one including the carnivorous forms, and *Sauropoda* and *Predentata*, for the two herbivorous groups, the last order being made up of three separate suborders; namely, the Stegosauria, the Ceratopsia, and the typical Ornithopoda. The first of these suborders contains large dinosaurs more or less protected by a dermal covering of bony plates; the second group includes the huge horned dinosaurs; and the third is made up of the forms that in shape and structure most nearly resemble birds.

143

The nearer relations of these groups to one another and to allied forms will be treated more fully in the concluding portion of this article.

The geological range of the Dinosauria, so far as at present known, is confined entirely to the Mesozoic, or the Age of Reptiles. The first indications of the group are found in the lower Triassic, and during this period these reptiles increased in number and size. In Jurassic time they were especially abundant, and in size and diversity of form far surpassed all other forms of vertebrate life then existing. During the entire Cretaceous they were represented by many strange and highly specialized types, and at the close of this period all apparently became extinct.

The wide geographical extent of these reptiles is also of interest. While North America seems to have contained the greatest number of different types, some of the larger species are now known to have lived in the southern half of this continent. Europe stands next to America in variety and number of these reptiles, large and small. In Asia, Africa, and Australia, also, characteristic remains have been discovered, and doubtless many more will be found at no distant day. The geological horizons in which the dinosaurian remains of the Old World occur are essentially the same as those in which the corresponding types have been found in America.

The introduction and succession of the Dinosauria in North America form a most interesting chapter in the life-history of this continent, and one that has an important bearing on geology as well. As these reptiles were the dominant types of land animals during the whole of Mesozoic time, and the circumstances under which they lived were especially favorable to the preservation of their remains, the latter mark definite geological horizons, which have proved of great service in ascertaining the age of large series of strata containing few other characteristic fossils. In this way one important horizon in the Jurassic and another in the Cretaceous have been accurately determined by the remains of the gigantic dinosaurs entombed in them, while still other lines have been approximately drawn by less characteristic fossils from the same group of reptiles.

In describing briefly the various dinosaurs now known to have lived in North America, it will be most instructive to begin with the oldest, in the Triassic, and then treat of their successors as they left their remains in subsequent deposits of Jurassic and Cretaceous age. To make this succession clear to the reader, the diagram on page 145 (fig. 1) has been prepared. This diagram represents the principal geological horizons of vertebrate fossils in North America, as determined by the writer, and if carefully examined will be found in reality to be a synopsis of the whole subject. The first appearance, so far as known, of each important group of vertebrate animals may be ascertained, approximately, from the data given. Some of the more recent genera of each group are also recorded, with the period in which they lived.

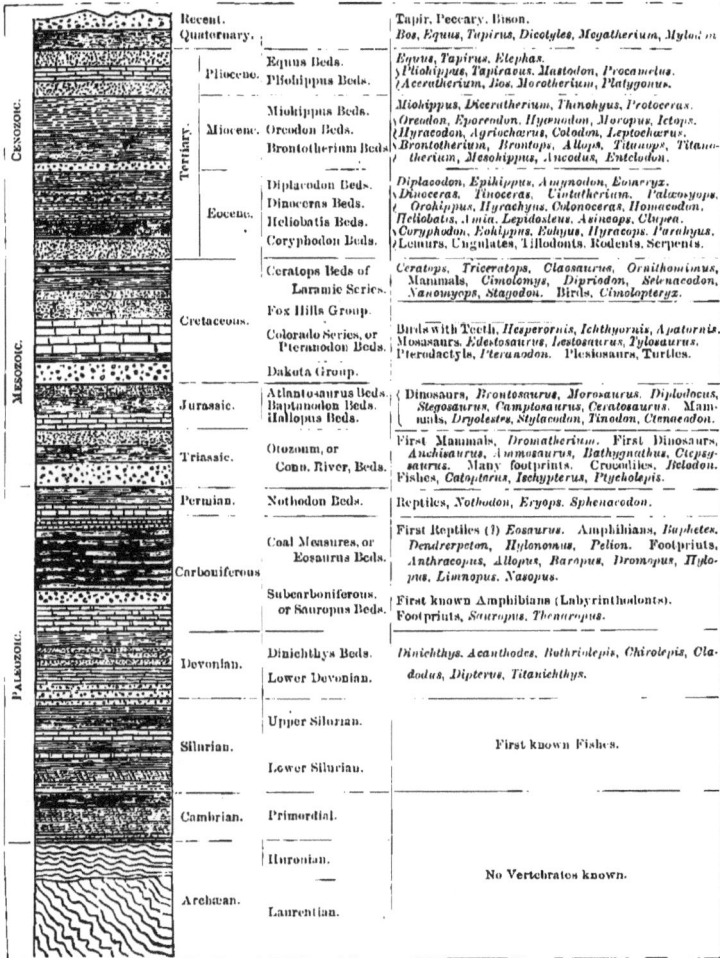

Era	Period	Beds	Vertebrate Fossils
CENOZOIC		Recent. Quaternary.	Tapir, Peccary, Bison. *Bos, Equus, Tapirus, Dicotyles, Megatherium, Mylodon*
	Pliocene.	Equus Beds. Pliohippus Beds.	*Equus, Tapirus, Elephas.* Pliohippus, Tapiravus, Mastodon, Procamelus. Aceratherium, Bos, Morotherium, Platygonus.
	Miocene.	Miohippus Beds. Oreodon Beds. Brontotherium Beds	*Miohippus, Diceratherium, Thinohyus, Protoceras.* Oreodon, Eporeodon, Hyænodon, Moropus, Ictops. Hyracodon, Agriochærus, Colodon, Leptochærus. Brontotherium, Brontops, Allops, Titanops, Titanotherium, Mesohippus, Ancodus, Entelodon.
	Eocene.	Diplacodon Beds. Dinoceras Beds. Heliobatis Beds. Coryphodon Beds.	*Diplacodon, Epihippus, Amynodon, Eohæryx.* Dinoceras, Tinoceras, Uintatherium, Palæosyops. Orohippus, Hyrachyus, Colonoceras, Homacodon. Heliobatis, Amia, Lepidosteus, Asineops, Clupea. Coryphodon, Eohippus, Eohyus, Hyracops, Parahyus. Lemurs, Ungulates, Tillodonts, Rodents, Serpents.
MESOZOIC	Cretaceous.	Ceratops Beds of Laramie Series.	*Ceratops, Triceratops, Claosaurus, Ornithomimus,* Mammals, Cimolomys, Dipriodon, Selenacodon, Nanomyops, Stagodon. Birds, Cimolopteryx.
		Fox Hills Group. Colorado Series, or Pteranodon Beds. Dakota Group.	Birds with Teeth, *Hesperornis, Ichthyornis, Apatornis.* Mosasaurs. *Edestosaurus, Læstosaurus, Tylosaurus.* Pterodactyls, *Pteranodon.* Plesiosaurs, Turtles.
	Jurassic.	Atlantosaurus Beds. Baptanodon Beds. Hallopus Beds.	Dinosaurs, *Brontosaurus, Morosaurus, Diplodocus,* Stegosaurus, Camptosaurus, Ceratosaurus. Mammals, Dryolestes, Stylacodon, Tinodon, Ctenacodon.
	Triassic.	Otozoum, or Conn. River, Beds.	First Mammals, *Dromatherium.* First Dinosaurs, Anchisaurus, Ammosaurus, Bathygnathus, Clepsysaurus. Many footprints. Crocodiles. *Belodon.* Fishes, *Catopterus, Ischypterus, Ptycholepis.*
PALÆOZOIC	Permian.	Nothodon Beds.	Reptiles, *Nothodon, Eryops. Sphenacodon.*
	Carboniferous.	Coal Measures, or Eosaurus Beds.	First Reptiles (?) *Eosaurus.* Amphibians, *Baphetes,* Dendrerpeton, Hylonomus, Pelion. Footprints, Anthracopus, Allopus, Baropus, Dromopus, Hylopus, Limnopus, Nasopus.
		Subcarboniferous, or Sauropus Beds.	First known Amphibians (Labyrinthodonts). Footprints, *Sauropus, Thenaropus.*
	Devonian.	Dinichthys Beds. Lower Devonian.	*Dinichthys, Acanthodes, Bothriolepis, Chirolepis, Cladodus, Dipterus, Titanichthys.*
	Silurian.	Upper Silurian. Lower Silurian.	First known Fishes.
	Cambrian.	Primordial.	
		Huronian.	No Vertebrates known.
	Archæan.	Laurentian.	

FIG. 1.—GEOLOGICAL HORIZONS OF VERTEBRATE FOSSILS IN NORTH AMERICA.

PART I.

TRIASSIC DINOSAURS.

THEROPODA.

The remains of dinosaurs first discovered in this country were found in the Triassic sandstone of the Connecticut Valley, so famous for its fossil footprints, many of which were long supposed to have been made by birds. It is a remarkable fact that the first discovery in this sandstone was that of the skeleton of a true dinosaur, found in East Windsor, Conn., in 1818, many years before the first footprints were recorded. This discovery was announced in the American Journal of Science for November, 1820, and later numbers contain descriptions of the remains, some of which are now preserved in the museum of Yale University.

FIG. 2.—Slab of Connecticut River sandstone; showing footprints of two dinosaurs on a surface marked by raindrop impressions. One-tenth natural size. Triassic, Massachusetts.

When the footprints in the Connecticut sandstone first attracted attention, in 1835, many of these impressions resembled so closely those made by birds that they were from the first attributed to that class, and for many years it was not seriously questioned that all the three-toed impressions, even the most gigantic, were really the footprints of birds. The literature on this subject is very extensive, but its value to science has been seriously impaired by the discovery of dinosaurian remains in various parts of the world, which prove that many of these reptiles were remarkably bird-like and that their tracks could not be distinguished from those of birds.

It was also found that some of the most bird-like footprints of the
Connecticut Valley were not made by birds, but by quadrupeds which
usually walked on their hind feet, yet sometimes put their fore feet
to the ground. Others occasionally sat down, and left an impression
which proved that they, too, were not birds. Still others showed reptilian affinities in various ways; so that to-day it may be stated that
there is no evidence that any of these impressions in the Connecticut
sandstone were made by birds. This is true, also, of similar bird-like
footprints from strata of the same age in different portions of this
country, and will likewise hold good for similar impressions from other
parts of the world. It is quite probable that birds existed during the
Triassic period, but at present there is no proof of it.

ANCHISAURIDÆ.

A few bones of a dinosaur were found at Upper Milford, Lehigh
County, Pa., in 1847, in strata regarded as Triassic. The animal was
named *Clepsysaurus pennsylvanicus* by Dr. Isaac Lea, in the Proceed-
ings of the Academy of Natural Sciences of Philadelphia, in 1851, and
he subsequently described and figured the remains in the Journal of
the Academy in 1853. They are now preserved in the museum of that
society.

The next discovery of importance in this formation was reported
from Prince Edward Island, Canada. The specimen was an upper jaw
with teeth, in good preservation, indicating a true dinosaur of considerable size. This specimen was figured and described under the name
Bathygnathus borealis by Dr. Leidy in the Journal of the Academy of
Natural Sciences of Philadelphia for 1854, and is now in the museum
of that institution.

The next important discovery of a Triassic dinosaur in this country
was made in the Connecticut sandstone about 1856, at Springfield,
Mass., and portions of the skeleton are now preserved at Amherst
College. This animal was a true carnivorous dinosaur, very similar to
the first one described, and from essentially the same horizon. This
discovery was announced by Prof. Edward Hitchcock in 1858, in his
Ichnology of New England, and the remains were described and figured by Edward Hitchcock, jr., in 1865, in a supplement to the above
volume. The animal was then named *Megadactylus polyzelus*, and
its affinities have since been discussed by various authors.

ANCHISAURUS.

A discovery of greater interest was made in 1884, near Manchester,
Conn. The skeleton of another carnivorous dinosaur of larger size, but
nearly allied to the one last mentioned, was found in a coarse conglom-
erate, in essentially the same horizon of the Connecticut River sandstone. This skeleton was probably complete and in position when
discovered, but as its importance was not recognized at the time the
posterior portion only was saved, which was secured later by the writer

for the Yale University museum. This part consisted of the nearly
entire pelvic arch, with both hind limbs essentially complete and in the
position they were when the animal died. The remains preserved
indicate an animal about 6 or 8 feet in length, which was named *Anchisaurus major* by the writer, in the American Journal of Science for
April, 1889.[1] This generic title replaced Megadactylus, which was preoccupied. Subsequently, in 1891, this specimen was made the type of
the genus Ammosaurus.

A still more important discovery of another small dinosaur was made
later at the same locality, only a few feet distant from the spot where
the fossil last mentioned was entombed. This reptile, named *Anchisaurus colurus* by the writer, is one of the most perfect dinosaurs yet discovered in the Triassic. The skull and limbs and most of the skeleton
were in fair preservation, and in natural position, so that nearly all
the important points of the osseous structure can be determined with
certainty. Some of these are here placed on record as typical of the
group.

THE SKULL.

The skull was somewhat crushed and distorted, but its main features
are preserved. In Pl. II, fig. 1, a side view of this skull is given, one-
half natural size. One prominent feature shown in this view is the
bird-like character of the skull. The nasal aperture is small and well
forward. There is an antorbital opening and a very large orbit. The
latter is elongated-oval in outline. It is bounded in front by the prefrontal, above by the same bone and a small extent of the frontal, and
further back by the postfrontal. The postorbital completes the orbit
behind and the jugal closes it below. The supratemporal fossa is large
and somewhat triangular in outline. The infratemporal fossa is quite
large and is bounded below by a slender quadratojugal. The quadrate is
much inclined forward. The teeth are remarkable for the great number
in use at one time. Those of the upper jaw are inclined forward, while
those below are nearly vertical. The lower jaw has the same general
features as this part in the typical Theropoda.

In Pl. III, figs. 1 and 2, the same skull is shown, also one-half
natural size. The top of the skull, represented in fig. 1, is considerably broken, and this has made it difficult to trace the sutures, but
the general form and proportions of the upper surface are fairly represented. In fig. 2 only the back portion of the cranium is shown. The
foramen magnum is remarkably large, and the occipital condyle is
small and oblique. The basipterygoid processes are unusually short.

The neck vertebræ of this skeleton are long and slender and very
hollow. Their articular ends appear to be all plane or slightly concave. The trunk vertebræ are more robust, but their centra are quite
long. The sacrals appear to be three in number.

---

[1] The original descriptions of nearly all the other fossils discussed in the present paper may be found
in the same journal.

The scapular arch is well preserved. The scapula, shown in Pl. II, fig, 2, *s*, is very long, with its upper end obliquely truncated. The coracoid is unusually small and imperforate. The sternum was of cartilage, some of which is preserved. The humerus is of the same length as the scapula and its shaft is very hollow. The radius and ulna also are both hollow, and are nearly equal in size.

There is but one carpal bone ossified in this specimen, and this is below the ulna. There were five digits in the manus, but only three of functional importance, the first, second, and third, all armed with sharp claws. The fifth was quite rudimentary. The fore foot of the type species of Anchisaurus is shown, one-half natural size, on Pl. III, fig. 5.

The pelvic bones are shown in fig. 3 of Pl. II. The ilium is small, with a slender preacetabular process. The ischia are elongated, and their distal ends are slender and not expanded at the extremity. The pubes are also long, imperforate, and not coossified with each other. The anterior part is a plate of moderate width.

The femur is much curved and longer than the tibia. The latter is nearly straight, with a narrow shaft. The fibula when in position was not close to the tibia, but curved outward from it. All these bones have very thin walls. The astragalus is small, closely applied to the tibia, and has no ascending process. The calcaneum is of moderate size and free. There are only two tarsal bones in the second row.

The hind foot had four functional digits, all provided with claws. The fifth was represented only by a rudiment of the metatarsal. The first digit was so much shorter than either the second, third, or fourth, that this foot would have made a three-toed track very much like the supposed bird-tracks of the Connecticut River sandstone.

ANCHISAURUS SOLUS.

A fortunate discovery has recently brought to light almost the entire skeleton of still another diminutive dinosaur, which may be referred to Anchisaurus, but clearly belongs to a distinct species. It was found in nearly the same horizon as the remains above described, and in the immediate vicinity, so there can be little doubt that it was a contemporary. The skeleton is embedded in a very coarse matrix, so difficult to remove that the investigation is only in part completed. The portions uncovered show the animal to have been about 3 feet in length, and of very delicate proportions. The bones of the skeleton are nearly all extremely light and hollow, but most of them are in a fair state of preservation.

The skull, so far as it can now be observed, resembles the one just described. The teeth are numerous, and inclined forward. The orbit

is very large. The quadrate is inclined forward, and the lower jaw is
robust. The entire skull is about 65ᵐᵐ long, and the lower jaws are
of the same length.

The neck was very long and slender, the first five cervicals measuring 80ᵐᵐ in extent. The dorsals are also elongated, the last six covering a space of 135ᵐᵐ. The number of vertebræ in the sacrum can not
yet be determined. The caudal vertebræ are short, the first ten occu-
pying a space of 140ᵐᵐ.

The humerus has a very large radial crest, and is 66ᵐᵐ in length.
The rest of the fore limb, so far as made out, is similar to that in the
species described. The tibia is about 88ᵐᵐ in length. There were five
digits in the hind foot, but the fifth is represented only by the rudimentary metatarsal. The animal was about as large as a small fox.

AMMOSAURUS.

The genus Ammosaurus, represented by remains of larger size from
the same strata, was also a typical carnivorous dinosaur, and apparently a near ally of Anchisaurus. So far as at present known, the
footprints of the two reptiles would be very similar, differing mainly
in size.

On Pl. III, fig. 6, is shown an entire hind foot of Ammosaurus, one-
fourth natural size. In this foot the tarsus is more complete than in
Anchisaurus. The astragalus has no true ascending process, the calcaneum is closely applied to the end of the fibula, and there are three
well-developed bones in the second row. The fifth digit had only a
single phalanx. The sacrum and ilia of the type species of Ammosaurus are shown in fig. 3, and the ischia of Anchisaurus in fig. 4.

RESTORATION OF ANCHISAURUS.

PLATE IV.

The Triassic dinosaurs known from eastern North America have now
been briefly reviewed. Remains of seven individuals are sufficiently
well preserved to indicate the main characters of the animals to which
they pertained. These were all carnivorous forms of moderate size,
and the known remains are from essentially the same geological horizon.

The genus Anchisaurus, one of the oldest known members of the
Theropoda, is so well represented by parts of four skeletons from these
deposits that a restoration of one species can now be made with considerable certainty. This has been attempted, and the result is given,
one-twelfth natural size, in the accompanying Pl. IV. The animal was
about 6 feet in length.

The skeleton chosen for this restoration is the type specimen of
Anchisaurus colurus, described by the writer in the American Journal
of Science in 1891 and 1893. This skeleton when discovered was entire,
and apparently in the position in which the animal died. Portions of

the neck and the tail vertebræ were unfortunately lost, but the skull and nearly all the rest of the skeleton were saved. The parts missing are fortunately preserved in a smaller specimen of an allied species (*Anchisaurus solus*) found at the same locality, and these have been used to complete the outline of the restoration. Portions of two other specimens, nearly allied, and from the same horizon, were also available, and furnished some suggestions of value.

The restoration, as shown on Pl. IV, indicates that *Anchisaurus colurus* was one of the most slender and delicate dinosaurs yet discovered, being surpassed in this respect only by some of the smaller bird-like forms of the Jurassic. The position chosen is one that must have been habitually assumed by the animal during life, but the comparatively large fore limbs suggest the possibility of locomotion on all four feet. The compressed terminal digits of the fore feet, however, must have been covered by very sharp claws, which were used mainly for prehension, and not for walking.

The small head and bird-like neck are especially noticeable. The ribs of the neck and trunk are very slender. The tail apparently differed from that of any other dinosaur hitherto described, as it was evidently quite slender and flexible. The short neural spines and the diminutive chevrons, directed backward, indicate a tail not compressed, but nearly round, and one usually carried free from the ground.

DINOSAURIAN FOOTPRINTS.

The present restoration will tend to clear up one point long in doubt. The so-called " bird tracks" of the Connecticut River sandstone have been a fruitful subject of discussion for half a century or more. That some of these were not made by birds has already been demonstrated by finding with them the impressions of fore feet. Although no bones were found near them, others have been regarded as footprints of birds because it was supposed that birds alone could make such series of bipedal, three-toed tracks and leave no impression of a tail.

It is now evident, however, that a dinosaurian reptile like Anchisaurus and its near allies must have made footprints very similar to, if not identical with, the "bird tracks" of this horizon. On a firm but moist beach only three-toed impressions would have been left by the hind feet, and the tail could have been kept free from the ground. On a soft, muddy shore the claw of the first digit of the hind foot would have left its mark, and perhaps the tail also would have touched the ground. Such additional impressions the writer has observed in various series of typical "bird tracks" in the Connecticut sandstone, and all of them were probably made by dinosaurian reptiles. On Pl. V and also p. 146, fig. 2, are shown several series of Triassic footprints, which were probably all made by dinosaurs. No tracks of true birds are known in this horizon.

DISTRIBUTION OF TRIASSIC DINOSAURS.

It is a remarkable fact that the seven skeletons of Triassic dinosaurs now known from the eastern part of this continent are all carnivorous forms and of moderate size. There is abundant evidence from footprints that large herbivorous dinosaurs lived here at the same time, but no bones nor teeth have yet been found. In the western part of this country a few fragments of a large dinosaur have been discovered in strata of supposed Triassic age, but with this possible exception osseous remains of these forms appear to be wanting in this horizon.

Fragmentary remains, also, of dinosaurs have been found in the Triassic deposits of Pennsylvania and North Carolina, but they throw little light on the animals they represent. Footprints, apparently made by dinosaurs, occur in New Jersey in the same horizon as those of the Connecticut Valley. Impressions of similar form have been discovered also in the Triassic sandstones of New Mexico. A few bones of a large dinosaurian were found by Prof. J. S. Newberry, in strata apparently of this age, in southeastern Utah. These remains were named by Professor Cope, *Dystrophæus viæmalæ*, in 1877, but their near affinities have not been determined. A single vertebra, apparently belonging in this group, had been previously found at Bathurst Island, Arctic America, and described by Prof. Leith Adams, in 1875, under the generic name Arctosaurus.

The European Triassic dinosaurs, with which the American forms may be compared, are mainly represented by the two genera Thecodontosaurus Riley and Stutchbury, from the upper Trias, or Rhætic, near Bristol, in England, and Plateosaurus (Zanclodon) von Meyer, from nearly the same horizon in Germany. The writer has investigated with care the type specimens and nearly all the other known remains of these genera found at these localities.

Remains of dinosaurs have been found in Triassic strata, also, in India, in South Africa, and in Australia, but the specimens discovered were mostly fragmentary, and apparently indicate no new types.

PART II.

JURASSIC DINOSAURS.

During the Jurassic period the dinosaurs of North America attained remarkable development, and, as a group, appear to have reached their culmination. The Theropoda, or carnivorous forms, which were so abundant, though of moderate size, in the Triassic, were represented in the Jurassic by many and various forms; some were very minute, but others were of gigantic size and dominated all living creatures during this age. The herbivorous dinosaurs, however, were the most remarkable of all, some far surpassing in bulk any known land animals; others, also of huge dimensions and clad in coats of mail, assumed the most bizarre appearance; while others still, diminutive in size and of light and graceful form, were so much like birds that only a comparative anatomist, with well-preserved skeletons of both before him, could tell one from the other. In this case, at least, a single tooth or bone would not suffice, though a Cuvier sat in judgment.

In the western part of this country, especially in the Rocky Mountain region, vast numbers of dinosaurs lived and flourished during all Jurassic time. Their remains are so abundant, and so perfectly preserved in many localities, that those already obtained have furnished the basis for a classification of the whole group. This classification, first proposed by the writer in the American Journal of Science in 1881, and subsequently emended, may be appropriately used here in considering the American dinosaurs from this formation. It will be discussed more fully in the concluding part of the present paper.

THEROPODA.

Near the base of the Jurassic in the Rocky Mountain region an interesting geological horizon has been defined as the Hallopus beds, since here only remains of a remarkable dinosaurian, named by the writer *Hallopus victor*, have been found. The position of this horizon is shown in the diagram on page 145 (fig. 1). Another reptile, Nanosaurus, the most diminutive dinosaur known, occurs in the same strata. This horizon is believed to be lower than that of the Baptanodon beds, although the two have not been found together. The Hallopus beds now recognized are in Colorado, below the Atlantosaurus beds, but are apparently quite distinct from them.

HALLOPUS.

The type specimen of Hallopus, the only one known, is the greater part of the skeleton of an animal about the size of a rabbit. This was described by the writer in 1877, and referred to the Dinosauria. On further investigation it was found to be distinct from all the known members of that group, and in 1881 it was made the type of a new suborder, the Hallopoda. One of the most distinctive characters, which separated it widely from all known dinosaurs, was seen in the tarsus, which had the calcaneum much produced backward. This feature, in connection with the greatly elongated metatarsals, suggested the generic name Hallopus, or leaping foot.

The general structure of the pelvis, especially of the ilium and pubis, as well as the proportions of the entire hind limb, suggested an affinity with Compsognathus, from the Jurassic of Bavaria, and the writer, in his classification of the dinosaurs, in 1882, placed the Hallopoda next to the suborder Compsognatha, which belongs in the great group of carnivorous dinosaurs, the Theropoda.

The writer has since reexamined the type specimen and had various parts of it uncovered, so far as the hard matrix of red sandstone would permit. This has brought to light other portions of the skeleton, so that now many of the more important characters of the group can be determined with certainty.

FORE AND HIND LIMBS.

In its present condition the specimen shows both the fore and hind limbs in good preservation, portions of the scapular arch, and apparently the entire pelvis and sacrum, various vertebræ, ribs, and other parts of the skeleton. It is doubtful if any portions of the skull are sufficiently well preserved for determination. On Pl. VI are given outline restorations of the fore and hind limbs of this specimen.

The scapula is of moderate length, and its upper portion broad and thin. The humerus is slender, with a strong radial crest. The shaft is very hollow, with thin walls, and the cavity extends almost to the distal end. The latter is but little expanded transversely. The radius and ulna are short, and were closely applied to each other. There were but four digits in the manus, the first being short and stout, and the others slender.

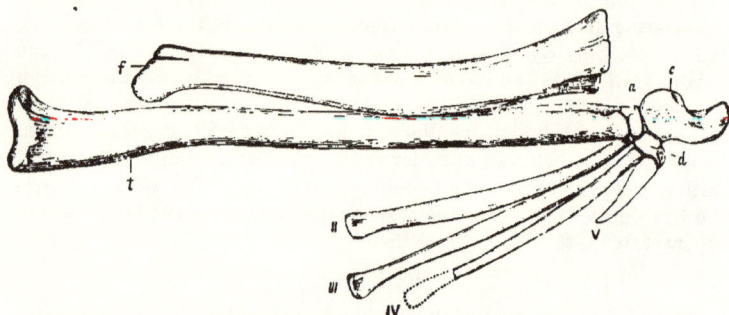

FIG. 3.—Left leg and foot of *Hallopus victor* Marsh; side view. Natural size.
a, astragalus; *c*, calcaneum; *d*, tarsal; *f*, femur; *t*, tibia; *II*, second metatarsal; *V*, remnant of fifth metatarsal.

All three pelvic bones aided in forming the acetabulum, as in typical dinosaurs. The ilia are of the carnivorous type, and resemble in form those of Megalosaurus. The pubes are rod-like, and projected downward and forward. The distal ends are closely applied to each other, but not materially expanded, and in the present specimen are not coossified with each other. The ischia projected downward and backward, and their distal extremities are expanded, somewhat as in the Crocodilia.

The femur is comparatively short, with the shaft curved and very hollow. The tibia is nearly straight, much longer than the femur, and its shaft equally hollow. The fibula was slender and complete, but tapered much from above downward. Its position was not in front of the tibia, as in all known dinosaurs, but its lower extremity was outside, and apparently somewhat behind, the tibia.

The astragalus is large, and covered the entire end of the tibia, but was not coossified with it. The calcaneum is compressed transversely, and much produced backward. It was closely applied to the outside of the astragalus, and although agreeing in general form with that of a crocodile, strongly resembles the corresponding bone in some mammals. The tarsal joint was below the astragalus and calcaneum. There appears to be but a single bone in the second tarsal row, although this may be composed of two or more elements.

There were but three functional digits in the hind foot, and their metatarsals are greatly elongated. The first digit seems to be wanting, and the fifth is represented only by a remnant of the metatarsal. The posterior limbs, as a whole, were especially adapted for leaping, and are more slender than in almost any other known reptile.

The main characters of the posterior limb are shown in fig. 3, on the opposite page, which represents the bones of the left leg and foot, natural size, in the position in which they lay when uncovered. All the bones figured are still firmly embedded in the matrix.

There are but two vertebræ in the sacrum. The other vertebræ preserved have their articular faces biconcave. The chevrons are slender and very elongate.

Taken together, the known characters of Hallopus clearly indicate dinosaurian affinities rather than those of any other group of reptiles, and if the Dinosauria are considered a subclass the Hallopoda at present may be regarded as a group of dinosaurs standing further apart from typical forms than any other.

CŒLURUS.

In the horizon above, the Atlantosaurus beds of the upper Jurassic, the carnivorous dinosaurs are of larger size, and some of them were among the most powerful and ferocious reptiles known. The one nearest Hallopus in size and general characteristics is Cœlurus, described by the writer in 1879 and now known from several skeletons, although no good skull has yet been discovered.

The skull of Cœlurus is known only from fragments. The teeth are typical of the order Theropoda. One is shown on Pl. VII, fig. 1. The most marked feature in all the known remains of Cœlurus is the extreme lightness of the bones, the excavations in them being more extensive than in the skeleton of any other known vertebrate. In the vertebræ, for example, the cavities are proportionately larger than in either pterodactyls or birds, the amount of osseous tissue retained being mainly confined to their exterior walls. In Pl. VII cervical, dorsal, and caudal vertebræ are figured, with transverse sections of each to illustrate this point. Even the ribs of Cœlurus are hollow, with well-defined walls to their large cavities.

THE VERTEBRÆ.

The vertebræ of Cœlurus now known are from various parts of the column, and most of them are in good condition. Three of these are represented, natural size, in Pl. VII. The cervicals are large and elongate, and were locked together by strong zygapophyses. The first three or four behind the axis had the front articular face of the centrum somewhat convex, and the posterior one deeply concave. All the other cervicals were biconcave, as were also the vertebræ of the trunk and tail. The articular faces of the cervicals are inclined, showing that the neck was curved. The anterior cervical ribs were coossified with the centra, as in birds. Figs. 2, 2a, and 2b, Pl. VII, represent a cervical vertebra from near the middle of the neck. The cavities in the cervicals are connected with the outside by comparatively large pneumatic openings. The neural canal is very large, and traces of the neurocentral suture are distinct.

The dorsal vertebræ of Cœlurus are much shorter than the cervicals. The centra have a deep cup in front and a shallow concavity behind. These articular faces are nearly at right angles to the axis of the trunk. The neural spine is elevated and compressed. The transverse processes are elongate. The ribs preserved have undivided heads. A posterior dorsal is represented in Pl. VII, figs. 3, 3a, and 3b. The suture of the neural arch is distinct in this specimen. The foramina leading to the cavities in the dorsal vertebræ are quite small.

The caudal vertebræ are elongate and very numerous. They are all biconcave, and all appear to have been without chevron bones. An anterior caudal is figured in Pl. VII, fig. 4, and the accompanying section shows the inner structure. In most of the caudals, the neurocentral suture has entirely disappeared.

THE HIND LIMBS.

The limb bones of Cœlurus are very hollow, and some of them appear pneumatic. The metatarsals are quite long and slender. The terminal phalanges of the hind feet are pointed, and in life were evidently covered with sharp claws. The ilium is of the Megalosaurus type. The pubes are slender, strongly coossified with each other, and terminated below by a large foot-like process, as shown in Pl. X, figs. 3 and 4.

The characters of Cœlurus are so distinctive that it appears to represent a separate family, which has been called by the writer the Cœluridæ. Several species of the genus are known in this country, all of moderate dimensions, varying in size from that of a fox to that of a wolf. Nearly all are from the Atlantosaurus beds of the West, but one small species has been found in the Potomac beds on the Atlantic Coast.

CERATOSAURUS.

The most interesting carnivorous dinosaur from the American Jurassic, and the one best known, is Ceratosaurus, which differs so widely from the typical forms that it has been regarded as representing a distinct suborder. The type specimen of Ceratosaurus, described by the writer in 1884, presented several characters not before seen in the Dinosauria. One of these is a horn core on the skull; another is a new type of vertebra as strange as it is unexpected; and a third is seen in the pelvis, which has the bones all coossified, as in existing birds. Among adult birds Archæopteryx alone has the pelvic bones separate, and this specimen of Ceratosaurus is the first dinosaur found with all the pelvic bones anchylosed. The metatarsal bones are also coossified, a feature characteristic of birds, but not known hitherto in any dinosaur.

THE SKULL.

The skull of *Ceratosaurus nasicornis* is very large in proportion to the rest of the skeleton. The posterior region is elevated, and moderately expanded transversely. The facial portion is elongate, and tapers gradually to the muzzle. Seen from above, the skull resembles in general outline that of an alligator. The nasal openings are separate and lateral, and are placed near the end of the snout, as shown in Pl. VIII.

Seen from the side, this skull appears lacertilian in type, the general structure being light and open. From this point of view one special feature of the skull is the elevated, trenchant horn core, situated on the nasals (Pl. VIII, fig. 1, *b*). Another feature is the large openings on the side of the skull, four in number. The first of these is the anterior nasal orifice; the second, the very large triangular antorbital foramen; the third, the large oval orbit; and the fourth, the still larger lower temporal opening. A fifth aperture, shown in the top view of the skull (Pl. VIII, fig. 3, *h*), is the supratemporal fossa. These openings are all characteristic of the Theropoda, and are found also in the Sauropoda, but the antorbital foramen is not known in any other Dinosauria.

The plane of the occiput, as bounded laterally by the quadrates, slopes backward. The quadrates are strongly inclined backward, thus forming a marked contrast to the corresponding bones in Diplodocus and other Sauropoda. The occipital condyle is hemispherical in general form, and is somewhat inclined downward, making a slight angle with the long axis of the skull. The basioccipital processes are short and stout. The paroccipital processes are elongate and flattened, and but little expanded at their extremities. They extend outward and downward, to join the head of the quadrate.

The hyoid bones appear to be four in number. They are elongate, rod-like bones, somewhat curved, and in the present specimen were found nearly in their original position.

The parietal bones are of moderate size, and there is no pineal foramen. The median suture between the parietals is obliterated, but that between these bones and the frontals is distinct.

The frontal bones are of moderate length, and are closely united on the median line, the suture being obliterated. Their union with the nasals is apparent on close inspection.

The nasal bones are more elongate than the frontals, and the suture uniting the two moieties is obsolete. These bones support entirely the large, compressed, elevated horn core on the median line. The lateral surface of this elevation is very rugose, and furrowed with vascular grooves. It evidently supported a high, trenchant, horn, which must have formed a most powerful weapon for offense and defense. No similar weapon is known in any of the carnivorous Dinosauria, but it is not certain whether this feature pertained to all the members of this group or was only a sexual character.

The premaxillaries are separate, and each contained only three functional teeth. In the genera Compsognathus and Megalosaurus, of this order, each premaxillary contained four teeth, the same number found in the Sauropoda. In the genus Creosaurus, from the American Jurassic, the premaxillaries each contain five teeth, as shown in Pl. XII, fig. 1.

The maxillary bones in the present specimen are large and massive, as shown in Pl. VIII, fig. 1. They unite in front with the premaxillaries by an open suture; with the nasals, laterally, by a close union; and with the jugal behind, by squamosal suture. The maxillaries are each provided with fifteen functional teeth, which are large, powerful, and trenchant, indicating clearly the ferocious character of the animal. These teeth have the same general form as those of Megalosaurus, and the dental succession appears to be quite the same.

Above the antorbital foramen on either side is a high elevation composed of the prefrontal bones. These protuberances would be of service in protecting the orbit, which they partially overhang.

The orbit is of moderate size, oval in outline, with the apex below. It is bounded in front by the lachrymal, above this by the prefrontal, and at the summit the frontal forms for a short distance the orbital border. The postfrontal bounds the orbit behind, but the jugal completes the outline below.

The jugal bone is ⊥-shaped, the upper branch joining the postfrontal, the anterior branch uniting with the lachrymal above and with the maxillary below. The posterior branch passes beneath the quadratojugal, and with that bone completes the lower temporal arch, which is present in all known dinosaurs.

The quadratojugal is an L-shaped bone, and its anterior branch is united with the jugal by a close suture. The vertical branch is closely joined to the outer face of the quadrate.

The quadrate is very long and compressed antero-posteriorly. The head is of moderate size, and is inclosed in the squamosal. The lower extremity of the quadrate has a double articular face, as in some birds. One peculiar feature of the quadrate is a strong hook on the upper half of the outer surface. Into this hook a peculiar process of the quadratojugal is inserted, as shown in Pl. VIII, fig. 1.

The pterygoid bones are very large and extend well forward. The posterior extremity is applied closely to the inner side of the quadrate. The middle part forms a pocket, into which the lower extremity of the basipterygoid process is inserted. To the lower margin of the pterygoid is united the strong, curved transverse bone which projects downward below the border of the upper jaws, as shown in Pl. VIII, fig. 1, t.

There is a very short, thin columella, which below is closely united to the pterygoid by suture, and above fits into a small depression of the postfrontal.

The palatine bones are well developed, and after joining the pterygoids extend forward to the union with the vomers. The latter are apparently of moderate size. .

The parasphenoid is well developed and has a long, pointed, anterior extremity.

The whole palate is remarkably open, and the principal bones composing it stand nearly vertical, as in the Sauropoda.

THE BRAIN.

The brain in Ceratosaurus was of medium size, but comparatively much larger than in the herbivorous dinosaurs. It was quite elongate, and situated somewhat obliquely in the cranium, the posterior end being inclined downward. The position of the brain in the skull, and its relative size, are shown in Pl. VIII, fig. 3. A side view of the brain cast is shown in Pl. LXXVII, fig. 2.

The foramen magnum is small. The cerebellum was of moderate size. The optic lobes were well developed and proportionately larger than the hemispheres. The olfactory lobes were large and expanded. The pituitary body appears to have been of good size.

THE LOWER JAWS.

The lower jaws of Ceratosaurus are large and powerful, especially in the posterior part. In front the rami are much compressed, and they were joined together by cartilage only, as in all dinosaurs. There is a large foramen in the jaw, similar to that in the crocodile, as shown in Pl. VIII, fig. 1, f'. The dentary bone extends back to the middle of this foramen. The splenial is large, extending from the foramen forward to the symphysial surface, and forming in this region a border to the upper margin of the dentary. There were fifteen teeth in each ramus, similar in form to those of the upper jaws.

THE VERTEBRÆ.

The cervical vertebræ of Ceratosaurus differ in type from those in all other known reptiles. With the exception of the atlas, which is figured in Pl. IX, fig. 1, all are strongly cupped on the posterior end of each centrum. In place of an equally developed ball on the anterior end, there is a perfectly flat surface. The size of the latter is such that it can be inserted only a short distance in the adjoining cup, and this distance is accurately marked on the centrum by a narrow articular border, just back of the flat, anterior face. This peculiar articulation leaves more than three-fourths of the cup unoccupied by the succeeding vertebra, forming, apparently, a weak joint. This feature is shown in Pl. IX, figs. 2, 3, and 4.

The discovery of this new form of vertebra shows that the terms opisthocœlian and procœlian, in general use to describe the centra of vertebræ, are inadequate, since they relate to one end only, the other being supposed to correspond in form. The terms convexo-concave, concavo-convex, plano-concave, etc., would be more accurate and equally euphonious.

In Ceratosaurus, as in all the Theropoda except Cœlurus, the cervical ribs are articulated to the centra, not coossified with them, as in the Sauropoda. The latter order stands almost alone among dinosaurs in this respect, as all the Predentata—Stegosauria, Ceratopsia, and the Ornithopoda—have free ribs in the cervical region.

The dorsal and lumbar vertebræ are biconcave, with only moderate concavities. The sides and lower surface of the centra are deeply excavated, except at the ends, as shown in Pl. IX, fig. 5. These vertebræ show the diplosphenal articulation seen in Megalosaurus, and also in Creosaurus, as shown in Pl. XII, fig. 5.

All the presacral vertebræ are very hollow, and this is also true of the anterior caudals.

There are five well-coossified vertebræ in the sacrum in the present specimen of *Ceratosaurus nasicornis*. The transverse processes are very short, each supported by two vertebræ, and they do not meet at their distal ends.

The caudal vertebræ are biconcave. All the anterior caudals, except the first, supported very long chevrons, indicating a high, thin tail, well adapted to swimming (Pl. IX, fig. 6). The tail was quite long, and the distal caudals were very short.

THE SCAPULAR ARCH.

The scapular arch of Ceratosaurus is of moderate size, but the fore limbs were very small. The humerus is short, with a strong radial crest. The radius and ulna are also very short, and nearly equal in size. The carpal bones were only imperfectly ossified. There were four digits in the fore foot, and all were armed with sharp claws. The second and third digits were much larger than the first and fourth, and the fifth was entirely wanting.

THE PELVIC ARCH.

The pelvic bones in the Theropoda have been more generally misunderstood than any other portion of the skeleton in dinosaurs. The ilia, long considered as coracoids, have been usually reversed in position; the ischia have been regarded as pubes; while the pubes themselves have not been considered as part of the pelvic arch.

Fortunately, in the present specimen of Ceratosaurus, the ilium, ischium, and pubes are firmly coossified, so that their identification and relative positions can not be called in question. The ilia, moreover, were attached to the sacrum, which was in its natural place in the skeleton, and the latter was found nearly in the position in which the animal died. The pelves of Ceratosaurus and of Allosaurus are shown in Pl. X.

The ilium in Ceratosaurus has the same general form as in Megalosaurus. In most of the other Theropoda, also, this bone has essentially the same shape, and this type may be regarded as characteristic of the order, except in Triassic forms. In Creosaurus the anterior wing is more elevated, and the emargination below it wider, as shown in Pl. XII, fig. 2, but this may be due in part to the imperfection of the border.

The ischia in Ceratosaurus are comparatively slender. They project well backward, and for the last half of their length the two are in close apposition. The distal ends are coossified and expanded, as shown in Pl. X.

The pubes in Ceratosaurus have their distal ends coossified, as in all Jurassic Theropoda except Hallopus. They project downward and forward, and their position in the pelvis is shown in Pl. X. Seen from the front, they form a Y-shaped figure, which varies in form in different genera. The upper end joins the ilium by a large surface, and the ischium by a smaller attachment. The united distal ends are expanded into an elongate, massive foot, as shown in Pl. X, which is one of the most peculiar and characteristic parts of the skeleton. The pubes of Cœlurus are represented on the same plate.

The extreme narrowness of the pelvis is one of the most marked features in this entire group, being in striking contrast to its width in the herbivorous forms found with them. If the Theropoda were viviparous, which some known facts seem to indicate, one difficulty, naturally suggested in the case of a reptile, is removed.

Another interesting point is the use of the large foot at the lower end of the pubes, which is the most massive part of the skeleton. The only probable use is that it served to support the body in sitting down. That some Triassic dinosaurs sat down on their ischia is proved conclusively by the impressions in the Connecticut River sandstone. In such cases the leg was bent so as to bring the heel to the ground. The same action in the present group would bring the foot of the pubes to the ground, nearly or quite under the center of gravity of the animal. The legs and ischia would then naturally aid in keeping the body balanced. Possibly this position was assumed habitually by these ferocious biped reptiles while lying in wait for prey.

The femur is much curved, and the shaft very hollow. The tibia is shorter than the femur, nearly straight, and has a large cnemial crest. The astragalus is not coossified with the femur, and has a strong ascending process. The fibula is well developed, and nearly straight, its distal end fitting into the calcaneum. The tarsals of the second row are very thin, and united to the metatarsals below them.

One of the most interesting features in the extremities of Ceratosaurus is seen in the metatarsal bones, which are completely anchylosed, as were the bones of the pelvis. There are only three metatarsal elements in the foot, the first and fifth having apparently disappeared entirely. The three metatarsals remaining, which are the second, third, and fourth, are proportionately shorter and more robust than in the other known members of the order Theropoda, and, being firmly united to each other, they furnish the basis for a very strong hind foot.

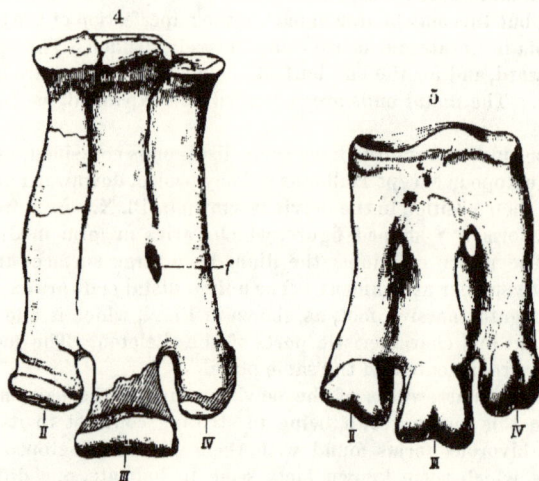

FIG. 4.—United metatarsal bones of *Ceratosaurus nasicornis* Marsh; left foot; front view. One-fourth natural size.

FIG. 5.—United metatarsal bones of great Penguin (*Aptenodytes Pennanti* G. R. Gr.); left foot; front view. Natural size.

f, foramen; *II, III IV*, second, third, and fourth metatarsals.

In fig. 4, above, these coossified metatarsals of Ceratosaurus are represented, and in fig. 5 the corresponding bone of a penguin is given for comparison.

In comparing these two figures, it will be seen that the three metatarsal elements of the dinosaur are quite as closely united as those of the bird. To the anatomist familiar with the tarsometatarsal bones of existing birds the specimen represented in fig. 4 will appear even more like this part in the typical birds than the one shown in fig. 5.

The position of the foramen, as seen in fig. 4, *f*, is especially characteristic of recent birds, and, as a whole, the hind foot of this Jurassic dinosaur was evidently similar to that of a typical bird.

All known adult birds, living and extinct, with possibly the single exception of Archæopteryx, have the metatarsal bones firmly united, while all the Dinosauria, except Ceratosaurus, have these bones separate. The exception in each case brings the two classes near together at this point, and their close affinity has now been clearly demonstrated.

RESTORATION OF CERATOSAURUS.

PLATE XIV.

The restoration of Ceratosaurus on Pl. XIV represents the reptile one-thirtieth natural size, and in a position it must have frequently assumed.

ALLOSAURUS.

Of the other carnivorous dinosaurs of the American Jurassic, three genera, Allosaurus, Creosaurus, and Labrosaurus, are especially worthy of notice. All were represented by species of large size, the natural enemies of the gigantic herbivorous forms that were so abundant in the same period. All had powerful jaws, sharp, cutting teeth, and a flexible neck. The fore limbs were quite small, and the feet were armed with strong claws for seizing living prey. The hind limbs were large and strong, and the animals used them alone in ordinary locomotion. These three genera may be separated by distinctive characters, and it is probable that they were not all contemporaneous.

The genus Allosaurus contains the largest carnivorous dinosaurs known. It may be readily distinguished from Ceratosaurus by the vertebræ and the pelvis, or the feet.[1] The cervicals are opisthocœlian instead of plano-concave, and the pelvic bones and metatarsals are free, as shown in Pls. X and XI. In Creosaurus, a smaller allied form, the teeth in the premaxillaries are more numerous, while the sacrum contains fewer vertebræ (Pl. XII). Labrosaurus is evidently a quite different type, for the dentary bone is edentulous in front, as shown in Pl. XIII.

EUROPEAN THEROPODA.

From the Jurassic of Europe the best-known carnivorous form is Megalosaurus, so named by Buckland, in 1824, the type specimen having been found in England, near Oxford. Although the first genus of dinosaurs described, but little has been made out in regard to the structure of the skull, and many portions of the skeleton remain to be determined. Its nearest American representative is probably Allosaurus, and both genera include species of large size.

The most interesting member of the Theropoda known in Europe is the diminutive specimen described by Wagner, in 1861, as *Compsognathus longipes*. The type specimen, the only one known, is from the lithographic slates of Solenhofen, Bavaria, and is now preserved in the museum in Munich. Fortunately, the skull and nearly all the skeleton are preserved, and as the specimen has been studied by many anatomists its more important characters have been made out. It is regarded as representing a distinct suborder, and no nearly related forms are known in Europe. A restoration in outline of this interesting dinosaur has been prepared by the writer, and will be found on Pl. LXXXII.

[1] The skull of *Allosaurus ferox* Marsh has an aperture in the maxillary in front of the antorbital opening. This aperture is not present in Ceratosaurus.

SAUROPODA.

The herbivorous dinosaurs of the American Jurassic are of special interest. To begin with the order Sauropoda, which includes the most primitive and gigantic forms, it is an interesting fact that the first specimen found in this country was one of the rarest of the group, and one of the most diminutive. A few teeth and bones only were obtained by Prof. P. T. Tyson, about 1858, near Bladensburg, Md. The teeth were named Astrodon by Dr. Christopher Johnston, in 1859, and in 1865 were described and figured by Dr. Leidy. The type specimens are now in the Yale museum, and one tooth is represented below in fig. 6. The strata containing these remains are known as the Potomac beds, but their exact age is a matter of doubt. They have been referred by some geologists to the Jurassic, and by others to the Cretaceous.

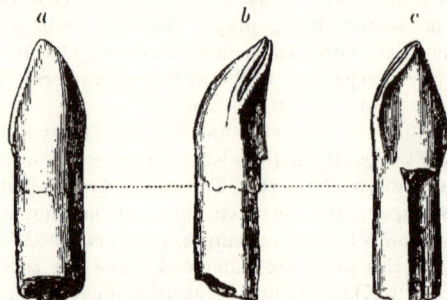

Fig. 6.—Tooth of *Astrodon Johnstoni* Leidy. Natural size. Potomac, Maryland. a, outer view; b, end view; c, inner view.

ATLANTOSAURUS BEDS.

The first known specimen of Sauropoda from the West was secured by the writer in August, 1868, near Lake Como, in Wyoming Territory. This fossil, an imperfect vertebra belonging to the genus since named Morosaurus, was found in the upper Jurassic clays, in the horizon now known as the Atlantosaurus beds. The section on page 145 will show the position of these beds in the geological scale, and their relation to other deposits in which Dinosauria have been found. This locality has since become one of the most famous in the entire Rocky Mountain region, and the writer has secured from it remains of several hundred dinosaurs, among which are many of the type specimens here described.

Remains of an enormous dinosaurian were found in 1877, near Morrison, Colo., by Prof. Arthur Lakes and Capt. H. C. Beckwith, U. S. N., and this was the beginning of a series of similar discoveries. These remains, described by the writer in the American Journal of Science for July of that year, proved to be those of a dinosaur far surpassing in size any previously known, and having characters that indicated a new order of these reptiles.

When first found these fossils were supposed to be from the Dakota group, but their upper Jurassic age was soon after determined by the writer from evidence that placed the horizon beyond dispute. The name *Titanosaurus montanus* was given by the writer to this reptile when first described, but as the generic designation proved to be preoccupied, Atlantosaurus was substituted.

A third Rocky Mountain locality which proved to be especially prolific in dinosaurs was found the following year, by Mr. M. P. Felch, a few miles north of Canyon, Colo., and in essentially the same horizon as the last-mentioned locality. Here were found the type specimens of some of the most interesting dinosaurs yet discovered in this country, all of them in fine preservation, and not infrequently in the exact position in which they died. Other localities of interest have been found in the same region.

Another locality of Sauropoda, more recently explored by the writer, is in South Dakota, on the eastern slope of the Black Hills. This is the most northern limit now known of the Atlantosaurus beds, which form a distinct horizon along the eastern flanks of the Rocky Mountains, marked at many points by the bones of gigantic dinosaurs, for nearly 500 miles. The strata are mainly shales or sandstones of fresh-water or estuary origin. They usually rest unconformably upon the red Triassic series, and have above them the characteristic Dakota sandstones.

On the western slope of the Rocky Mountains the Atlantosaurus beds are also well developed, especially in Wyoming, but here they have immediately below them a series of marine strata, which the writer has named the Baptanodon beds, from the largest reptile found in them. This horizon, also of Jurassic age, is shown in the section on page 145. One of the best exposures of the entire Jurassic series may be seen near Lake Como, Wyoming.

Besides the dinosaurs, which are especially abundant, the Atlantosaurus beds also contain numerous remains of extinct crocodiles, tortoises, and fishes, and with them have been found a small pterodactyl and a single bird. Many small mammals, also, have been described by the writer from the same beds.

FAMILIES OF SAUROPODA.

The Sauropoda of the American Jurassic are worthy of special attention, and so far as now known they may be divided into four families: the Atlantosauridæ, which include the largest forms; the Diplodocidæ and Morosauridæ, both represented by gigantic species; and the small Pleurocœlidæ, which were apparently the last survivors of the order in this country. Remains of the first three families are abundant in the Atlantosaurus beds of the West, but it is not certain that all were contemporaneous. The Pleurocœlidæ are especially characteristic of the Potomac beds on the Atlantic coast.

All the known members of these families were quadrupedal, with the fore and hind limbs nearly equal in length. The head was very small and the neck long, with its vertebræ opisthocœlian and lightened by inner cavities, thus allowing free motion.

The limb bones of all were solid, and the feet plantigrade, with five toes on each. The tail was especially long and massive. The general form and proportions of these reptiles are indicated in Pl. XLII, which represents the skeleton of a species of Brontosaurus, one of the typical genera of the Atlantosauridæ.

ATLANTOSAURIDÆ

ATLANTOSAURUS.

The present family was named by the writer in 1877, the type genus being Atlantosaurus. The type specimen of the first species described, *Atlantosaurus montanus*, is the sacrum represented in fig. 1, Pl. XVII, which shows characteristic features of the sacrum of the entire group Sauropoda, and thus distinguishes it from that of the other known dinosaurs. A second and larger species, *Atlantosaurus immanis*, was described by the writer in the following year, and on Pl. XVI are represented two pelvic bones and a femur, which belong to the type specimen, and give an idea of its gigantic size. The femur is over 6 feet long, and this, with other portions of the skeleton, indicates an animal about 70 or 80 feet in length. The pubis and ischium, represented in position in fig. 1, are especially characteristic of the family, as will be seen by comparing them with the corresponding parts of other allied genera, as shown in Pl. XXXVI.

At the same locality where these remains were found, portions of a skull were discovered, one of which is figured on Pl. XV. This specimen, which is the posterior part of the skull, is of much interest, and shows characters which separate it from all other corresponding remains of dinosaurs. The most marked feature is a distinct pituitary canal leading from the brain cavity down through the base of the skull, as shown in fig. 2 of Pl. XV. This canal appears to be a marked character of the family Atlantosauridæ. Other points of interest in these remains will be discussed later in the present paper.

APATOSAURUS.

Another genus of the present family is Apatosaurus, also described by the writer in 1877, and from the same geological horizon in Colorado. The sacrum represented in fig. 2, Pl. XVII, may be regarded as the type specimen. It has the same general features as the sacrum of Atlantosaurus, shown on that plate, but it has only three coossified vertebræ instead of four.

THE SACRAL CAVITY.

The neural canal in this sacrum, and indeed in all the sacra of the Sauropoda, is much enlarged, being especially expanded above each vertebral centrum, thus leaving a vaulted chamber in the united neural arches of the sacral vertebræ. A cast of this cavity in the type specimen of Apatosaurus is shown in fig. 3, Pl. XVIII.

This enlargement of the neural cord in the sacral region exists to some degree in reptiles and birds now living, but does not approach that found in the Sauropoda, or especially that in the Stegosauria, where, as will be shown later in the present article, this expansion reaches its maximum, and its functional importance must make it a dominant factor in the movements of the reptiles in which it is so highly developed. This great development has been found only in extinct reptiles in which the brain was especially diminutive, and the relation of the two nervous centers to each other offers a most interesting problem to physiologists.

THE VERTEBRÆ.

In Pl. XVIII, fig. 1, is shown a posterior cervical vertebra of Apatosaurus, and in fig. 2 of the same plate a dorsal vertebra is also represented, both being typical of the family Atlantosauridæ. The cervical

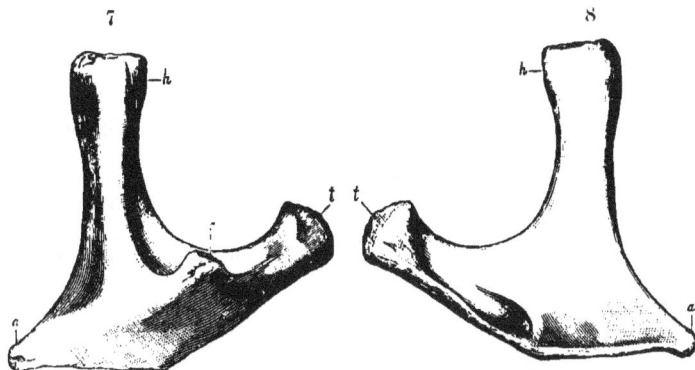

FIG. 7.—Cervical rib of *Apatosaurus ajax* Marsh; outer view.
FIG. 8.—The same rib; inner view.
Both figures are one-eighth natural size. *a*, anterior extremity; *h*, head; *r*, posterior process; *t*, tubercle.

vertebra, seen from behind, shows the deep, transverse cup of the posterior articular end of the centrum, as well as the coossified cervical ribs, both typical of the Sauropoda. A cervical rib of one species is shown in figs. 7 and 8.

The dorsal vertebra, seen from in front, presents the convex anterior ball of the centrum, and also the massive neural arch of the vertebra, with its elevated metapophyses, constituting a neural spine. The expanded diapophyses, or transverse processes, are especially noteworthy, as they aid in supporting the massive ribs, their extremities articulating with the tubercle of the rib, while the head is supported at the base of the arch by a sessile facet representing the parapophysis of the cervicals. The small neural canal in each vertebra is also an interesting feature, especially when contrasted with the expanded cavity in the sacrum shown in fig. 3 of the same plate.

The scapula and coracoid of Apatosaurus, shown in Pl. XIX, fig. 1, are also characteristic features of the family they represent. The shaft of the scapula is not expanded above in any of the genera of the Atlantosauridæ, although this expansion is characteristic of the genus Morosaurus and the family Morosauridæ, as shown in the same plate, fig. 2. Another important difference is indicated on this plate, in the coracoids, that of Apatosaurus being nearly square in outline, while in Morosaurus and its allies the contour of the coracoid is subovate.

BRONTOSAURUS.

The best-known genus of the Atlantosauridæ is Brontosaurus, described by the writer in 1879, the type specimen being a nearly entire skeleton, by far the most complete of any of the Sauropoda yet discovered. It was found in the Atlantosaurus beds, near Lake Como, Wyoming, and the remains were nearly in the position in which they were left at the death of the animal. This fortunate discovery has done much to clear up many doubtful points in the structure of the whole group Sauropoda, and the species *Brontosaurus excelsus* may be taken as a typical form, especially of the family Atlantosauridæ. The animal was about 60 feet in length. A second species, equally gigantic, has since been found in the same region.

In Pl. XX, fig. 1, a characteristic tooth of Brontosaurus is shown, which may also be regarded as typical for the family. In fig. 2 of the same plate the dentary bone is shown, with the teeth in outline. This bone is one of the most characteristic of the whole skeleton, as will be seen by comparing it with the corresponding parts of other Sauropoda represented in the following plates.

The genus Brontosaurus may be readily distinguished from all the other Sauropoda by the sacrum, which is composed of five anchylosed vertebræ, none of the other genera in this group having more than four. The sternum, moreover, consists of two separate bones, which are parial, and were united to each other on the median line apparently by cartilage only. In several other respects the genus resembles Morosaurus.

The present species, aside from its immense size, is distinguished by the peculiar lightness of its vertebral column, the cervical, dorsal, and sacral vertebræ all having very large cavities in their centra. The first three caudals, also, are lightened by excavations in their sides, a feature first seen in this genus, and one not observed in the other families of this group.

THE SCAPULAR ARCH.

The scapular arch in the present species is, fortunately, better known than that of any other member of the present order. In Pl. XXII the various bones are represented in position, and in fig. 2 of the same plate a sternal bone is shown separately. The scapula resembles in general form the corresponding bone in Apatosaurus, but the shaft is longer and the upper end somewhat wider.

The coracoid approaches more nearly that of Apatosaurus, which is subquadrate in outline. In Pl. XXII the scapula and coracoid of the present species are placed nearly in the same plane, and the space between them probably represents about the amount of cartilage which originally separated them. Both scapulæ were found in apposition with their respective coracoids.

The two sternal bones lay side by side between the two coracoids, and in this plate they are represented nearly as found. They are sub-oval in outline, concave above, and convex below. They are parial, and when in position nearly or quite meet on the median line. Each bone is considerably thickened in front, and shows a distinct facet for union with the coracoid. The posterior end is thin and irregular. The sternal ribs represented in figs. 12–15, p. 171, were found near the sternal bones. The sternum of a young ostrich is shown for comparison on Pl. XXII, fig. 3.

THE CERVICAL VERTEBRÆ.

The cervical vertebræ of the present species are quite numerous, thirteen at least belonging in this part of the column. All are strongly opisthocœlian. The anterior cervicals are very small in comparison with those near the dorsal region. From the third vertebra to the middle of the neck the centra increase in length and especially in bulk, but the posterior cervicals gradually become shorter. In Pl. XX, figs. 3 and 4, the sixth cervical is represented, and this is typical for the anterior half of the neck. All the anterior cervicals have coossified ribs, as in birds. In the posterior cervicals the ribs occasionally become free (Pl. XXI, fig. 1). The articular facet for the head of the rib rises gradually on the side of the centrum, the tubercular articulation remaining on the diapophysis. None of the cervicals have a neural spine. The neural canal is comparatively small. The centra of all the cervicals have deep excavations in the sides, and the transverse processes are more or less cavernous. The posterior cervicals which bear free ribs are remarkable for the great size of the zygapophyses, which are here much larger than elsewhere in the series. The anterior cervicals have several lateral cavities, while those farther back have only one large foramen in each side of the centrum, as in the dorsals.

THE DORSAL VERTEBRÆ.

The dorsal vertebræ of this species have short centra, more or less opisthocœlian. There is a very large cavity in each side, which is separated from the one opposite by a thin vertical partition. The neural canal is much larger than in the cervicals. The anterior dorsals are distinctly opisthocœlian. The neural spine has no prominence in this region, but rises rapidly farther back. In Pl. XXI, figs. 2 and 3, a posterior dorsal is represented, which shows the peculiar character of the vertebræ in this part of the series. The neural spine is greatly developed and has its summit transversely expanded. The vertebræ in this region, as in all the known Sauropoda, have the peculiar diplosphenal articulation. This is shown in fig. 3. In the vertebra figured, at the base of the neural spine, there is a strong anterior projection, which was inserted into the cavity between and above the posterior zyga-pophyses of the vertebra in front. There appear to be no true lumbar vertebræ, as those near the sacrum supported free ribs of moderate size. The vertebræ in this region have both faces of the centrum nearly flat or biconcave. An anterior dorsal rib is shown below.

THE SACRUM.

The sacrum in the present species consists of five well-coossified vertebræ, and in the type specimen the centrum of the last lumbar is firmly united with it, as shown in Pl. XXIII. The striking feature

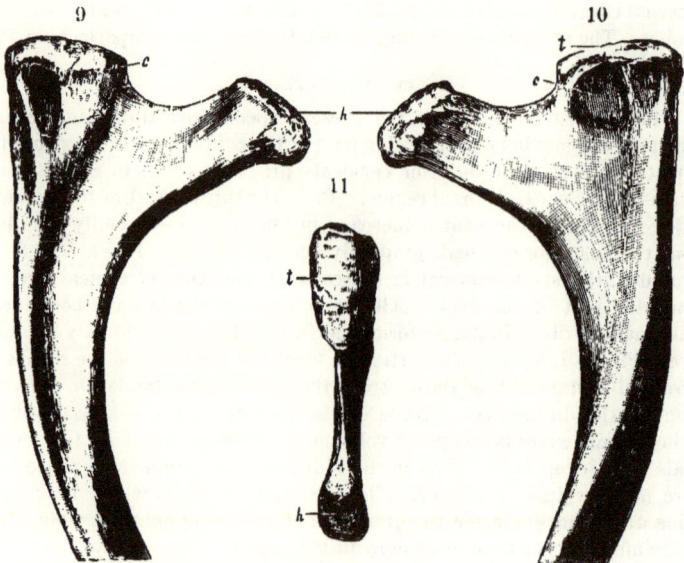

FIG. 9.—Proximal end of rib of *Brontosaurus excelsus* Marsh; front view.
FIG. 10.—The same bone; back view.
FIG. 11.—The same; superior view.
All the figures are one-eighth natural size. *c*, cavity; *h*, head; *t*, tubercle.

about this sacrum is the large general cavity it contained. This was divided in part by a median longitudinal partition, as shown in Pl. XXIII, fig. 2. The septum, however, was not continuous the whole length of the sacrum, so that the two lateral cavities were virtually one. This extended even into the lateral processes. The transverse partitions formed by the ends of the respective centra were also perforate, so that the sacrum proper was essentially a hollow cylinder. The cavernous character of the sacrum is one of the peculiar features of the suborder Sauropoda, and was described by the writer when the first species of this group was discovered in this country. The statement that any of the species has the sacrum solid is evidently based on erroneous observation.

Another peculiar character of the sacrum in the present genus is its lofty neural spine. This is a thin, vertical plate of bone with a thick massive summit, evidently formed by the union of the spines of several vertebræ. In front it shows rugosities for the ligament uniting it to the adjoining vertebra, and its posterior margin likewise indicates a similar union with the first caudal. In this genus, as in all the Sauropoda, each vertebra of the sacrum supports its own transverse processes. As shown in Pl. XXIII, the articulation for the ilium is formed by the coossification of the distal ends of the transverse processes. The neural canal is much enlarged in the sacrum, but less proportionally than in Stegosaurus.

14   15

12   13

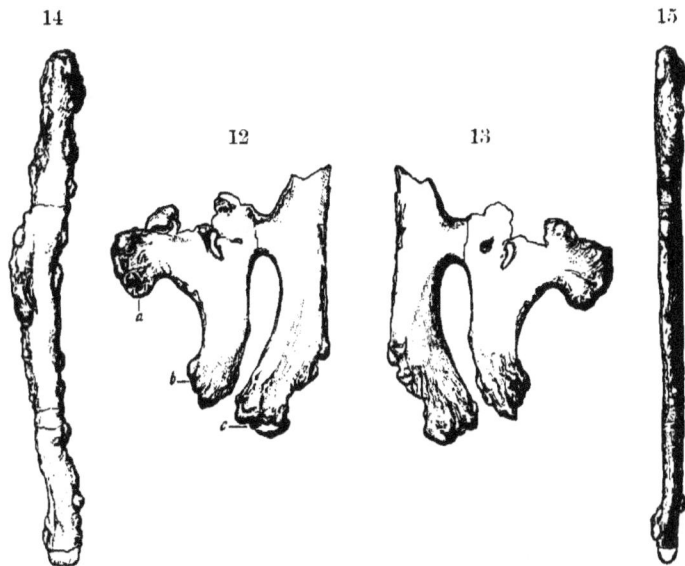

Fig. 12.—Sternal ribs of *Brontosaurus excelsus;* outer view.
Fig. 13.—The same specimen; inner view.
Fig. 14.—Sternal rib of same individual; outer view.
Fig. 15.—The same rib; inner view.
All the figures are one-eighth natural size.

THE CAUDAL VERTEBRÆ.

In the present species the three vertebræ next behind the sacrum have moderate-sized cavities between the base of the neural arch and the transverse processes. These shallow pockets extend into the base of the processes, but the centra proper are solid. All the other caudals have the centra, processes, and spines composed of dense bone. The fourth caudal vertebra, represented in Pl. XXIV, figs. 2 and 3, is

solid throughout, and the same is true of the chevron, figs. 4 and 5.
The neural spines of the anterior caudal vertebræ are elevated and
massive. The summit is cruciform in outline, due to the four strong
buttresses which unite to form it.

The median caudals all have low, weak spines, and no transverse
processes. The posterior caudals are elongate and without spines or
zygapophyses.

THE PELVIC ARCH.

The pelvic bones in the present species are shown in fig. 16. The
ilium represented is not quite perfect on its upper margin. Its anterior process for the support of the pubis is much larger than the poste-

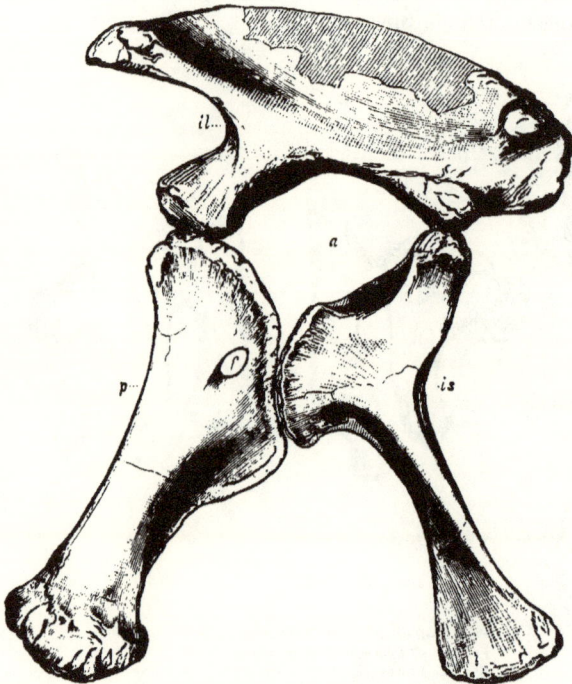

Fig. 16.—Pelvis of *Brontosaurus excelsus;* seen from the left. One-sixteenth natural size.
a, acetabulum; *f*, foramen in pubis; *il*, ilium; *is*, ischium; *p*, pubis.

rior one which meets the ischium. The pubis is elongate and massive.
It sends down a strong wing for union with the ischium, and has in
front of this the usual foramen. The distal end is expanded, and has
on the inner surface a rugose facet for union with its fellow by cartilage.
The ischium is more slender than the pubis, and has its lower end
expanded for symphysial union with the one on the other side (Pl.
XXIV, figs. 1 and 1*a*). This pelvis is more like that of Atlantosaurus

than any other of the known genera of the Sauropoda. The three bones shown in fig. 16 were found nearly in the position represented.

THE FORE LIMBS.

The fore limbs of Brontosaurus, as in most of the Sauropoda, were of large size and of massive proportions. The limb bones are all solid, and those of the feet are quite robust. There were five well-developed digits in the manus, and the metacarpals were all moderately elongate. A characteristic example is shown in figs. 17–20, below.

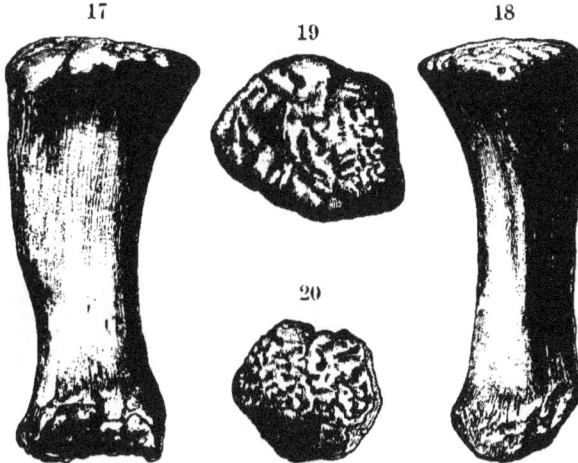

FIG. 17.—First metacarpal of *Brontosaurus amplus* Marsh; front view.
FIG. 18.—The same bone; side view.
FIG. 19.—Proximal end of same.
FIG. 20.—Distal end of same.
All the figures are one-fourth natural size.

THE HIND LIMBS.

The hind limbs of Brontosaurus were larger than those in front, and the bones were all solid, thus being in remarkable contrast to the elements of the vertebral column. The hind feet were plantigrade, and had five powerful digits. The first was very stout, and its terminal phalanx, shown in figs. 21–23, supported a powerful claw.

RESTORATION OF BRONTOSAURUS.

PLATE XLII.

Nearly all the bones represented in this restoration belonged to a single individual, which when alive was nearly or quite 60 feet in length. The position here given was mainly determined by a careful adjustment of these remains. That the animal at times assumed a position more erect than here represented is probable, but locomotion on the posterior limbs alone was hardly possible.

The head was remarkably small. The neck was long and flexible, and, considering its proportions, was the lightest portion of the vertebral column. The body was short, and the abdominal cavity of moderate size. The legs and feet were massive and the bones all solid. The feet were plantigrade, and each footprint must have been about a square yard in extent. The tail was large and nearly all the bones are solid.

The diminutive head will first attract attention, as it is smaller in proportion to the body than in any vertebrate hitherto known. The entire skull is less in diameter or actual weight than the fourth or fifth cervical vertebra.

A careful estimate of the size of Brontosaurus, as here restored, shows that when living the animal must have weighed more than 20 tons. The very small head and brain, and the slender neural cord, indicate a stupid, slow-moving reptile. The beast was wholly without offensive or defensive weapons or dermal armature.

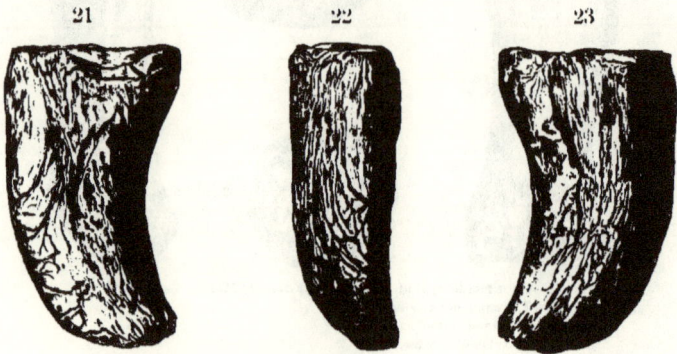

FIG. 21.—Terminal phalanx of *Brontosaurus excelsus;* outer view.
FIG. 22.—The same bone; front view.
FIG. 23.—The same; inner view.
All the figures are one-fourth natural size.

In habits Brontosaurus was more or less amphibious, and its food was probably aquatic plants or other succulent vegetation. The remains are usually found in localities where the animals seem to have been mired. The type specimen was discovered by W. H. Reed, near Lake Como, Wyoming.

· BAROSAURUS.

Another genus of the Sauropoda is indicated by various remains of a gigantic reptile described in 1890 by the writer. The most characteristic portions examined are the caudal vertebræ, which in general form resemble those of Diplodocus. They are concave below, as in the caudals of that genus, but the sides of the centra are also deeply excavated.

In the anterior caudals this excavation extends nearly or quite through the centra, a thin septum usually remaining. In the median caudals a deep cavity on each side exists, as shown in figs. 24–26, below.

On the distal caudals the lateral cavity has nearly or quite disappeared. All the caudal vertebræ are proportionally shorter than in Diplodocus, and their chevrons have no anterior projection, as in that genus.

The remains on which the present description is based are from the Atlantosaurus beds of South Dakota, about 200 miles farther north than this well-marked horizon has hitherto been recognized.[1]

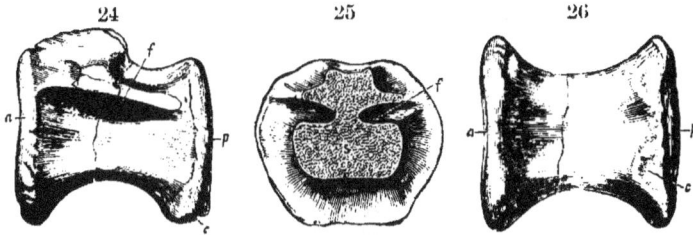

FIG. 24.—Caudal vertebra of *Barosaurus lentus* Marsh; side view.
FIG. 25.—The same vertebra, in section; front view.
FIG. 26.—The same vertebra; bottom view.

All the figures are one-eighth natural size. a. anterior end; c. face for chevron; f, lateral cavity; p, posterior end; s, section.

DIPLODOCIDÆ.

DIPLODOCUS.[2]

THE SKULL.

The skull of Diplodocus is of moderate size. The posterior region is elevated and narrow. The facial portion is elongate and the anterior part expanded transversely. The nasal opening is at the apex of the cranium, which from this point slopes backward to the occiput. In front of this aperture the elongated face slopes gradually downward to the end of the muzzle, as represented in Pl. XXV, fig. 1.

Seen from the side the skull of Diplodocus shows five openings: a small oval aperture in front, a large antorbital vacuity, the nasal aperture, the orbit, and the lower temporal opening. The first of these has not been seen in any other Sauropoda; the large antorbital vacuity is characteristic of the Theropoda also; while the other three openings are present in all the known Dinosauria.

On the median line, directly over the cerebral cavity of the brain, the type specimen of Diplodocus has also a fontanelle in the parietals. This, however, may be merely an individual peculiarity.

The plane of the occiput is of moderate size, and forms an obtuse angle with the frontoparietal surface.

The occipital condyle is hemispherical in form, and seen from behind is slightly subtrilobate in outline. It is placed nearly at right angles

---

[1] Strata that may represent this horizon have been observed still farther north, especially in Montana, but have not yet been identified by characteristic fossils.
[2] American Journal of Science, 1878–1884.

to the long axis of the skull. It is formed almost wholly of the basi-occipital, the exoccipitals entering but slightly or not at all into its composition. The basioccipital processes are large and rugose. The paroccipital processes are stout and somewhat expanded at their extremities, for union with the quadrates.

The parietal bones are small and composed mainly of the arched processes which join the squamosals. There is no true pineal foramen, but in the skull here figured (Pl. XXV) there is the small unossified tract mentioned above. In one specimen of Morosaurus a similar opening has been observed, but in other Sauropoda the parietal bones, even if thin, are complete. The suture between the parietals and frontal bones is obliterated in the present skull, and the union is firm in all the specimens observed.

The frontal bones in Diplodocus are more expanded transversely than in the other Sauropoda. They are thin along the median portion, but quite thick over the orbits.

The nasal bones are short and wide and the suture between them and the frontals is distinct. They form the posterior boundary of the large nasal opening, and also send forward a process to meet the ascending branch of the maxillary, thus taking part in the lateral border of the same aperture.

The nasal opening is very large, subcordate in outline, and is partially divided in front by slender posterior processes of the premaxillaries. It is situated at the apex of the skull, between the orbits, and very near the cavity for the olfactory lobes of the brain.

The premaxillaries are narrow below, and with the ascending processes very slender and elongate. Along the median line these processes form an obtuse ridge, and above they project into the nasal opening. Each premaxillary contains four functional teeth.

The maxillaries are very largely developed, more so than in most other known reptiles. The dentigerous portion is very high and slopes inward. The ascending process is very long, thin, and flattened, inclosing near its base an oval foramen, and leaving a large unossified space posteriorly. Above, it meets the nasal and prefrontal bones. Along its inner border for nearly its whole length it unites with the ascending process of the premaxillary. Each maxillary contains nine teeth, all situated in the anterior part of the bone (Pl. XXV, fig. 1).

Along their upper margin, on the inner surface, the maxillaries send off a thickened ridge, or process, which meets its fellow, thus excluding the premaxillaries from the palate, as shown in fig. 27, opposite. Above this, for a large part of their length, the ascending processes of the maxillaries underlap the ascending processes of the premaxillaries and join each other on the median line.

The orbits are situated posteriorly in the skull, being nearly over the articulation of the lower jaw. They are of medium size, nearly circular in outline, their plane looking outward and slightly backward. No indications of sclerotic plates have been found either in Diplodocus or in the other genera of Sauropoda.

The supratemporal fossa is small, oval in outline, and directed upward and outward. The lateral temporal fossa is elongated, and oblique in position, bounded, both above and below, by rather slender temporal bars.

The prefrontal and lachrymal bones are both small; the suture connecting them, and also that uniting the latter with the jugal, can not be determined with certainty.

The postfrontals are triradiate bones. The longest and most slender branch is that descending downward and forward for connection with the jugal; the shortest is the triangular projection directed backward and fitting into a groove of the squamosal; the anterior branch, which is thickened and rugose, forms part of the orbital border above.

The squamosal lies upon the upper border of the paroccipital process. The lower portion is thin and closely fitted over the head of the quadrate bone.

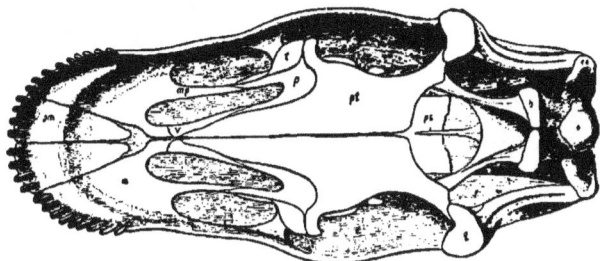

Fig. 27.—Skull of *Diplodocus longus* Marsh; seen from below. One-sixth natural size.
b, basioccipital process; *eo*, exoccipital; *m*, maxillary; *mp*, maxillary plate; *o*, occipital condyle; *p*, palatine; *pm*, premaxillary; *pt*, pterygoid; *ps*, parasphenoid; *q*, quadrate; *t*, transverse bone; *v*, vomer.

The quadrate is elongated and slender, with its lower end projecting very much forward. In front it has a thin plate extending inward and overlapping the posterior end of the pterygoid.

The quadratojugal is an elongate bone, firmly attached posteriorly to the quadrate by its expanded portion. In front of the quadrate it forms for a short distance a slender bar, which is the lower temporal arcade.

The palate is very high and roof-like, and composed chiefly of the pterygoids, as shown above in fig. 27. The basipterygoid processes are elongate, much more so than in the other genera of Sauropoda.

The pterygoids have a shallow cavity for the reception of these processes, but no distinct impression for a columella. Immediately in front of this cavity the pterygoids begin to expand, and soon form a broad, flat plate, which stands nearly vertical. Its upper border is thin, nearly straight, and extends far forward. The anterior end is acute and unites along its inferior border with the vomer. A little in front of the middle a process extends downward and outward, for union with the transverse bone. In front of this process, uniting with it and with the transverse bone, is the palatine.

The palatine is a small semioval bone fitting into the concave anterior border of the pterygoid, and sending forward a slender process for union with the small palatine process of the maxillary.

The vomer is a slender, triangular bone, united in front by its base to a stout process of the maxillary, which underlaps the ascending process of the premaxillary. Along its upper and inner border it unites with the pterygoid, except at the end, where for a short distance it joins a slender process from the palatine. Its lower border is wholly free.

THE BRAIN.

The brain of Diplodocus was very small, as in all dinosaurs from the Jurassic. It differed from the brain of the other members of the Sauropoda, and from that of all other known reptiles, in its position, which was not parallel with the longer axis of the skull, as is usually the case, but inclined to it, the front being much elevated, as in the ruminant mammals (Pl. LXXVI, fig. 4). Another peculiar feature of

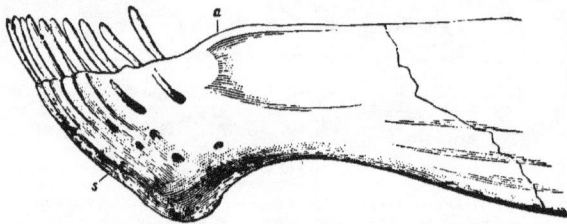

Fig. 28.—Dentary bone of *Diplodocus longus;* seen from the left. One-third natural size. *a,* edentulous border; *s,* symphysis.

the brain of Diplodocus was its very large pituitary body, inclosed in a capacious fossa below the main brain case. This character separates Diplodocus at once from the Atlantosauridæ, which have a wide pituitary canal connecting the brain cavity with the throat. In the Morosauridæ the pituitary fossa is quite small.

The posterior portion of the brain of Diplodocus was diminutive. The hemispheres were short and wide and more elevated than the optic region. The olfactory lobes were well developed, and separated in front by a vertical osseous septum. The very close proximity of the external nasal opening is a new feature in dinosaurs, and appears to be peculiar to the Sauropoda.

THE LOWER JAWS.

The lower jaws of Diplodocus are more slender than in any of the other Sauropoda. The dentary especially lacks the massive character seen in Morosaurus, and is much less robust than the corresponding

bone in Brontosaurus. The short dentigerous portion in front is de-
curved (Pl. XXV, fig. 1), and its greatest depth is at the symphysis, as
shown in fig. 28 above. The articular, angular, and surangular bones
are well developed, but the coronary and splenial appear to be small.

The dentition of Diplodocus is the weakest seen in any of the known
Dinosauria, and strongly suggests the probability that some of the more
specialized members of this great group were edentulous. The teeth
are entirely confined to the front of the jaws (Pl. XXV, fig. 1), and
those in use were inserted in such shallow sockets that they were readily
detached. Specimens in the Yale museum show that entire series of
upper or lower teeth could be separated from the bones supporting
them without losing their relative position. In Pl. XXVI, fig. 1, a
number of these detached teeth are shown.

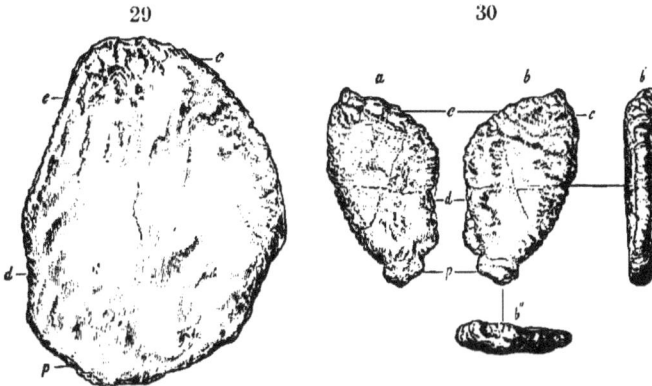

FIG. 29.—Sternal plate of *Brontosaurus amplus;* top view.
FIG. 30.—Sternal plate of *Morosaurus grandis* Marsh.
Both figures are one-eighth natural size. *a,* bottom view; *b,* top view; *b',* side view; *b'',* end view;
c, face for coracoid; *d,* margin next to median line; *e,* inner front margin; *p,* posterior end.

The teeth of Diplodocus are cylindrical in form and quite slender.
The crowns are more or less compressed transversely and are covered
with thin enamel, irregularly striated. The roots are long and slender
and the pulp cavity is continued nearly or quite to the crown. In the
type specimen of Diplodocus there are four teeth, the largest of the
series, in each premaxillary; nine in each maxillary, and ten in each
dentary of the lower jaws. There are no palatine teeth.

The jaws contain only a single row of teeth in actual use. These
are rapidly replaced, as they wear out or are lost, by a series of successional teeth, more numerous than is usual in these reptiles. Pl. XXVI,
fig. 2, represents a transverse section through the maxillary, just behind
the fourth tooth. The latter is shown in place, and below it is a series

of five immature teeth, in various stages of development, preparing to take its place. These successional teeth are lodged in a large cavity, which extends through the whole dental portion of the maxillary. The succession is also similar in the premaxillary teeth and in those of the lower jaws.

THE VERTEBRÆ.

The vertebral column of Diplodocus, so far as at present known, may be readily distinguished from that of the other Sauropoda by both the centra and chevrons of the caudals. The former are elongated and deeply excavated below, as shown in Pl. XXVI, figs. 4 and 5. The chevrons are especially characteristic, and to their peculiar form the generic name Diplodocus refers. They are double, having both anterior and posterior branches, and the typical forms are represented in figs. 6 and 7 of the above plate.

The cervical and dorsal vertebræ of Diplodocus are represented by typical examples on Pl. XXVI, fig. 3, and Pl. XXVII, and the sacrum with ilium attached is shown on Pl. XXVIII.

THE STERNAL BONES.

The sternal bones in Diplodocus are especially large, and in form resemble those in *Brontosaurus excelsus*. Those in *Brontosaurus amplus* are less robust, and are somewhat similar in shape to those of Morosaurus, as shown in figs. 29 and 30.

THE PELVIC GIRDLE.

A most characteristic bone of the two families of Sauropoda first described is the ischium. In the Atlantosauridæ the ischia are massive and directed downward, with their expanded extremities meeting on the median line. In the Morosauridæ the ischia are slender, with the shaft twisted about 90°, directed backward, and the sides meeting on the median line, thus approaching this part in the more specialized dinosaurs. The ischia referred to the genus Diplodocus (Pl. XXVIII, fig. 3) are intermediate in form and position between those above mentioned. The shaft is not expanded distally, nor twisted, but was directed downward and backward, with the sides meeting on the median line.

The feet of Diplodocus are shown in Pls. XXVIII and XXIX.

SIZE AND HABITS.

The type specimen of Diplodocus, to which the skull here figured belongs, indicates an animal intermediate in size between Atlantosaurus and Morosaurus, probably 40 or 50 feet in length when alive. The teeth show that it was herbivorous, and the food was probably succulent vegetation. The position of the external nares would seem to indicate in some measure an aquatic life.

MOROSAURIDÆ.

MOROSAURUS.

The genus Morosaurus, the type of the family, was described by the writer in 1878, in the American Journal of Science, which contains most of the original descriptions of Sauropoda found in this country.

THE SKULL.

The head in this genus was very small. The posterior part of the skull resembled that in Diplodocus, but the front was much more massive. The lower jaw was especially powerful, as shown by the dentary bone figured in Pl. XXX, fig. 3. This figure also shows the size and position of the teeth, one of which is figured in Pl. XXXI, figs. 1 and 2.

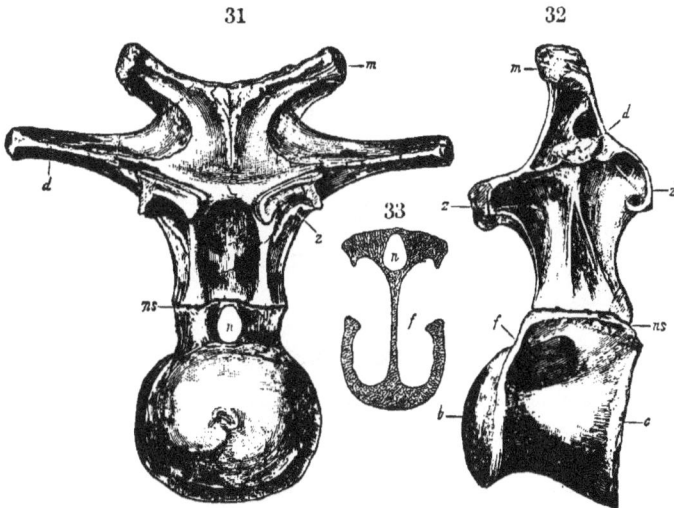

31   32

33

FIG. 31.—Anterior dorsal vertebra of *Morosaurus grandis;* front view.
FIG. 32.—The same vertebra; side view.
FIG. 33.—Transverse section through centrum of same.
All the figures are one-eighth natural size. *b,* ball ; *c,* cup ; *d,* diapophysis ; *f,* cavity in centrum ; *m* metapophysis ; *n,* neural canal ; *ns,* neural suture ; *z,* anterior zygapophysis ; *z',* posterior zygapophysis.

The brain was very small. Its form and position in the skull are shown in fig. 2 of Pl. XXX. At the back of the skull there are two peculiar bones, called by the writer the postoccipital bones, which are shown in Pl. XXX, fig. 1.

THE VERTEBRÆ.

The neck was elongated, and except the atlas all the cervical vertebræ have deep cavities in the sides of the centra, similar to those in birds of flight. They are also strongly opisthocœlous. The atlas and axis are not anchylosed together, and the elements of the atlas are separate (Pl. XXXI).

The dorsal vertebræ are distinctly opisthocœlous. The posterior dorsals have elongated transverse neural spines, and have deep cavities in the sides. An anterior dorsal is shown in figs. 31–33, p. 181. There are four vertebræ in the sacrum, all with cavities in the centra. Their transverse processes, or sacral ribs, are vertical plates with expanded ends. The anterior caudal vertebræ are plano-concave, and nearly or quite solid. The tail was elongated, and the chevrons are similar to those in crocodiles (Pl. XXXIX). The vertebræ of Morosaurus are represented on Pls. XXXI–XXXIV.

The scapula is elongated and very large, and the shaft has a prominent anterior projection. The coracoid is small, suboval in outline, and has the usual foramen near its upper border. These two bones are well represented in Pl. XIX, nearly in the relative position in which they were found. The humerus is very large and massive, and its radial crest prominent. This bone is nearly solid, and its ends were rough and well covered with cartilage. This is true, also, of all

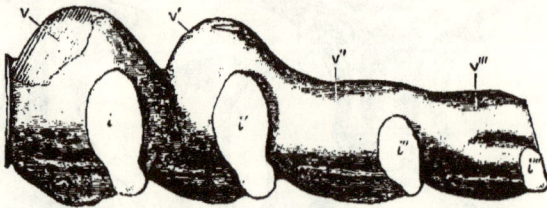

FIG 34.—Cast of neural cavity in sacrum of *Morosaurus lentus* Marsh; side view. One-fourth natural size.
i, i′, i″, i‴, intervertebral foramina; v, v′, v″, v‴, cavities in first, second, third, and fourth sacral vertebræ.

the large limb bones in this genus. The radius and ulna are nearly equal in size. The carpal bones are separate and quite short. The five metacarpals are elongated, and the first is the stoutest. The toes were thick, and the ungual phalanges were evidently covered with hoofs. In Pl. XXXVIII, fig. 1, the restoration of the scapular arch and entire fore limb of one species of Morosaurus well illustrates this part of the skeleton.

The pelvic bones are distinct from each other and from the sacrum. The ilium is short and massive, and shows on its inner side only slight indications of its attachment to the sacrum. More than half the acetabulum is formed by the ilium, which sends down in front a strong process for union with the pubis, and a smaller one behind to join the ischium (Pl. XXXV, fig. 1, a and b). The acetabulum is completed below by the pubis and ischium. The pubis is large and stout, and projects forward and downward, uniting with its fellow on the median line in a strong ventral symphysis. Its upper posterior margin meets the ischium and contains a large foramen. The ischium projects downward and backward, and its distal end is not expanded. The relative position and general form of the three pelvic bones in this genus are shown in Pl. XXXVI, fig. 1, and the entire pelvic arch in Pl. XXXI. The ischia of two species are shown on Pl. XXXV. The sacrum of one species of Morosaurus is shown on Pl. XXXI, and that of another on Pl. XXXIII. A cast of the sacral cavity of the latter is represented in fig. 34, p. 182.

THE HIND LIMBS.

The femur is long and massive, and without a true third trochanter, although a rugosity marks its position. The great trochanter is obtuse and is placed below the head. The ridge which plays between the tibia and fibula is distinct. The tibia is shorter than the femur. It is without a spine or fibular ridge, and its distal end shows that the astragalus was separated from it by a cushion of cartilage. The fibula is stout, its two extremities nearly equal, and its distal end supports the calcaneum. The tarsal elements of the second row are unossified. The five well-developed digits are similar to those in the manus. The first metatarsal is much the largest (Pl. XXIX, fig. 2). The feet are also shown on Pls. XXXVII and XXXVIII.

PLEUROCŒLIDÆ.

PLEUROCŒLUS. [1]

THE SKULL.

The genus Pleurocœlus includes the smallest individuals of the Sauropoda found in this country, most of them not being larger than existing crocodiles, and some even smaller. The skull is quite small, and resembles in its structure that of Morosaurus, but has points of similarity also with that of Brontosaurus. The teeth resemble those of Diplodocus most nearly, but have shorter roots and are much more numerous, the entire upper and lower jaws being furnished with teeth. The dentary bone is similar in shape and proportions to that of Brontosaurus, differing widely from that of Diplodocus and Morosaurus. It is slender and rounded at the symphysis, instead of having the massive, deep extremity seen in Morosaurus. The maxillary also is much less robust. The supraoccipital agrees closely in shape with that of Morosaurus, and forms the upper border of the foramen magnum, as in that genus. In Pl. XL, fig. 1, is shown the dentary of Pleurocœlus, with the teeth in outline, and beside it are typical examples of the teeth.

THE VERTEBRÆ.

The cervical vertebræ are very elongate and strongly opisthocœlons. The deep cavities in the sides of the centrum are separated only by a thin septum of bone, as shown in fig. 3 of the same plate. The dorsal

---

[1] American Journal of Science, 1888.

vertebræ are much longer than the corresponding vertebræ of Morosaurus, and have a very long, deep cavity in each side of the centrum,
to which the generic name refers. All the trunk vertebræ hitherto
found are proportionately nearly double the length of the corresponding
centra of Morosaurus, and the lateral cavity is still more elongate.
These points are shown in the posterior dorsal vertebra represented in
figs. 4 and 5 of Pl. XL. The neural arch in this region is lightened by
cavities, and is connected with that of the adjoining vertebræ by the
diplosphenal articulation. A dorsal centrum of another species is
shown below in figs. 35–37.

The sacral vertebræ in Pleurocœlus are more solid than in Morosaurus, but more elongate. The surface for the rib, or process which
abutsagainst the ilium, is well in front, more so than in any of the known

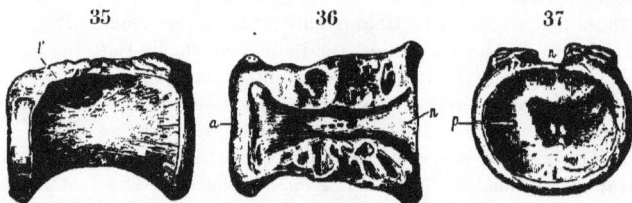

FIG. 35.—Dorsal vertebra of *Pleurocœlus montanus* Marsh; side view.
FIG. 36.—The same vertebra; top view.
FIG. 37.—The same; back view.

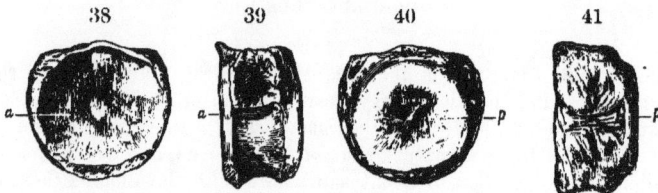

FIG. 38.—Caudal vertebra of same individual; front view.
FIG. 39.—The same vertebra; side view.
FIG. 40.—The same; back view.
FIG. 41.—The same; top view.
All the figures are one-half natural size. *a*, anterior end; *f*, cavity in centrum; *n*, neural canal; *p*, posterior end.

Sauropoda. Behind this articular surface is a deep pit, which somewhat
lightens the centrum. These characters are seen in the sacral vertebra
represented in figs. 6 and 7 of Pl. XL.

The first caudal vertebra has the centrum very short, and its two
articular faces nearly flat, instead of having the anterior surface deeply
concave, as in the other known Sauropoda. An anterior caudal is
shown in figs. 38–41, above. The neural spines in this region are compressed transversely. The middle and distal caudals are comparatively
short and the former have the neural arch on the front half of the
centrum, as shown in figs. 8 to 11 of Pl. XL.

The bones of the limbs and feet preserved agree in general with those of the smaller species of Morosaurus, but indicate an animal of slighter and more graceful build. The metapodials are much more slender and the phalanges are less robust than in the other members of the order, as shown in Pl. XLI.

DISTRIBUTION OF THE SAUROPODA.

In the preceding pages the most important forms of the Sauropoda now known from North America have been briefly described and illustrated. The only remains known from other parts of America are a few fragmentary specimens recently found in Patagonia, and described by Lydekker, under the generic names Argyrosaurus and Titanosaurus, in the Anales del Museo de La Plata, 1893. The specimens now known, although in poor preservation, show distinctive characters of the order Sauropoda, and indicate reptiles of gigantic size. The discovery is interesting and will doubtless soon be followed by others of more importance.

In England remains of Sauropoda have long been known, and the first generic name given was Cardiodon, proposed by Owen, in 1841; and based on teeth alone. A number of other generic names have since been proposed, and several are still in use. Among these are Cetiosaurus and Bothriospondylus Owen, Pelorosaurus Mantell, Æpysaurus and Hoplosaurus Gervais, Ornithopsis Seeley, and Eucamerotus Hulke. The absence of the skull, and the fact that most of the type specimens pertain to different parts of the skeleton, render it difficult, if not impossible, to determine the forms described, and especially their relations to one another.

COMPARISON WITH EUROPEAN FORMS.

In examining the European Sauropoda with much care the writer was impressed by three prominent features in the specimens investigated:

(1) The apparent absence of any characteristic remains of the Atlantosauridæ, which embrace the most gigantic of the American forms.

(2) The comparative abundance of another family, Cardiodontidæ, nearly allied to the Morosauridæ, but, as a rule, less specialized.

(3) The absence, apparently, of all remains of the Diplodocidæ.

A number of isolated teeth and a few vertebræ of one immature individual appear to be closely related to the Pleurocœlidæ, but this, for the present, must be left in doubt.

A striking difference between the Cardiodontidæ and the allied American forms is that in the former the fore and hind limbs appear to be more nearly of the same length, indicating a more primitive or generalized type. Nearly all the American Sauropoda, indeed, show a higher degree of specialization than those of Europe, both in this feature and in some other respects.

The identity of any of the generic forms of European Sauropoda with those of America is at present doubtful. In one or two instances it is impossible, from the remains now known, to separate closely allied forms from the two countries. Portions of one Wealden animal, referred by Mantell to Pelorosaurus, are certainly very similar to some of the smaller forms of Morosaurus, especially in the proportions of the fore limbs, which are unusually short. This fact would distinguish them at once from Pelorosaurus, and until the skull and more of the skeleton are known they can not be separated from Morosaurus.

The only Sauropoda reported from other parts of the world are some fragmentary remains from India, referred by Lydekker to the genus Titanosaurus, and more recently other remains from Madagascar, which the same author has placed in the genus Bothriospondylus. The wide distribution of the forms already known indicates that many future discoveries may be expected.

PREDENTATA.

The third order of the Dinosauria, according to the system of classification here adopted, is the one named by the writer the Predentata, a name derived from the fact that all the members of the group have a predentary bone, which is wanting in all other dinosaurs, and in fact in all other vertebrates, living and extinct. This order includes three suborders: the Stegosauria (plated lizards), the Ceratopsia (horned lizards), and the Ornithopoda (bird-footed). These are all herbivorous forms, and most of them contain species of very large size. The Stegosauria were mainly confined to the Jurassic, the Ceratopsia entirely to the Cretaceous, while the Ornithopoda were abundant in each of these periods. Of the Jurassic forms the Stegosauria will be first discussed, especially the typical family Stegosauridæ, which contains the American forms.

STEGOSAURIDÆ.

The genus Stegosaurus, the type of the family, was described by the writer in 1877 from a specimen found in the Atlantosaurus beds of Colorado. Subsequently other remains were discovered and described by the writer, the more important being from near Lake Como, Wyo., and Canyon, Colo., localities which have furnished so many type specimens of the Sauropoda and other dinosaurs.

STEGOSAURUS.

THE SKULL.

The skull of Stegosaurus is long and slender, the facial portion being especially produced. Seen from the side, with the lower jaw in position, it is wedge-shaped, with the point formed by the premaxillary, which projects well beyond the mandible, as shown in fig. 1, Pl. XLIII. The anterior nares are large and situated far in front. The orbit is very large and placed well back. The lower temporal fossa is somewhat smaller. All these openings are oval in outline and are on a line

nearly parallel with the top of the skull. In this view the lower jaw
covers the teeth entirely.

Seen from above, as shown in fig. 3, Pl. XLIII, the wedge-shaped form
of the skull is still apparent. The only openings visible are the supratemporal fossæ. The premaxillary bones are short above, but send back
a long process below the narial orifice. The nasal bones are very large
and elongate. They are separated in front by the premaxillaries, and
behind by anterior projections from the frontal bones. The prefrontals
are large, and are placed between the nasals and the prominent, rugose
supraorbitals. The frontals are short, and externally join the postfrontals. The parietals are small and closely coossified with each
other.

Viewed from in front, the skull and mandible present a nearly quadrate outline (Pl. XLIII, fig. 2), and the mutual relations of the facial
bones are well shown. In this view is seen, also, the predentary bone,
a characteristic feature of the mandible in this genus. The lateral
aspect of this bone is shown in fig. 1 of the same plate.

The teeth in this genus are entirely confined to the maxillary and
dentary bones, and are not visible in any of the figures here given.
They are small, with compressed, fluted crowns, which are separated
from the roots by a more or less distinct neck. The premaxillary and
the predentary bones are edentulous. A typical tooth of Stegosaurus,
and one of an allied genus, Priconodon, are represented on Pl. XLIV,
figs. 1 and 2.

The present skull belongs to the type specimen of a very distinct
species, *Stegosaurus stenops*. The skull and nearly complete skeleton of
this specimen, with most of the dermal armor in place, were found
almost in the position in which the animal died.

This reptile was much smaller than those representing the other
species of this genus. Its remains were found by Mr. M. P. Felch in
the Atlantosaurus beds of the Upper Jurassic, in southern Colorado.
In this geological horizon all known American forms of Stegosaurus
have been discovered.

THE BRAIN.

Little has been known hitherto of the brain in dinosaurs, but fortunately one specimen of Stegosaurus has the brain case well preserved
and apparently without distortion. Figs. 3 and 4 of Pl. XLIV show
the form and general characters of this brain cavity. The brain of
this reptile was much elongated, and its most striking features were
the large size of the optic lobes and the small cerebral hemispheres.
The latter had a transverse diameter only slightly in excess of the
medulla. The cerebellum was quite small. The optic nerve corre-
sponded in size with the optic lobes. The olfactory lobes were of large
size. As a whole, this brain was lacertilian rather than avian. A
brain cast of a young alligator is given on Pl. LXXVII for comparison.

The contrast in the development of the cerebral region is marked, but in some other respects the correspondence is noteworthy.

In comparing the proportionate size of the brain of this living reptile with that of Stegosaurus, as given on the same plate, the result proves of special interest. The absolute size of the two brain casts is approximately as 1 to 10, while the bulk of the entire bodies, estimated from corresponding portions of each skeleton, was as 1 to 1,000. It follows that the brain of Stegosaurus was only $\frac{1}{100}$ that of the alligator, if the weight of the entire animal is brought into the comparison. If the cerebral regions only of the two brains were compared the contrast would be still more striking. This comparison gives, of course, only approximate results, and some allowance should be made for the proportionally larger brain in small animals.

The brain of *Stegosaurus ungulatus* is clearly of a lower type than that of Morosaurus, which, as the writer has shown, was several times smaller in diameter than the neural canal in its own sacrum. In the latter genus the brain was proportionally shorter, and the cerebral region better developed, as shown in the plate cited. The absolute size of this brain as compared with that of Stegosaurus is about 16 to 10, the brain of the alligator figured being regarded as 1. Taking again the body of the alligator as the unit, and Stegosaurus as 1,000, that of Morosaurus would be about 1.500. Stegosaurus had thus one of the smallest brains of any known land vertebrate. These facts agree fully with the general law of brain growth in extinct mammals and birds as made out by the writer.[1]

THE ANTERIOR VERTEBRÆ.

The vertebræ of Stegosaurus preserved all have the articular faces of their centra concave, although in some the depression is slight. They are all, moreover, without pneumatic or medullary cavities. On Pl. XLV a selection from the vertebral series of one skeleton is given, which shows the principal forms. Figs. 1 and 2 represent a median cervical. The other neck vertebræ have their centra of similar length, but the diameter increases from the axis to the last of the series. Some of the anterior cervicals have a small tubercle in the center of each end of the centra, a feature seen also in some of the caudals. All the cervicals supported short ribs.

The dorsal vertebræ have their centra rather longer and more or less compressed. The neural arch is especially elevated. The neural canal is much higher than wide. The head of the rib fits into a pit on the side of the neural arch. Figs. 3 and 4 of Pl. XLV represent a posterior dorsal with characteristic features. The ribs are massive, and are strengthened by their form, which is T-shaped in transverse section.

[1] Odontornithes, a monograph on the extinct toothed birds of North America: U. S. Geol. Expl. Fortieth Par., Vol. VII, pp. 8, 121, 1880   Dinocerata, a monograph of an extinct order of gigantic mammals: Mon. U. S. Geol. Survey, Vol. X, Author's Edition, pp. 57–59, 1884.

THE SACRUM.

The true sacrum of Stegosaurus is composed of four well-coossified vertebræ. In fully adult animals the pelvic arch may be strengthened by the addition of one or more lumbar vertebræ, as in the specimen figured in Pl. XLVI, fig. 5, where two are firmly consolidated with the sacrum. The centra of the sacral vertebræ are solid, like the others in the column. Their neural arches are especially massive, and the spines have high and expanded summits. The transverse processes of the sacral vertebræ are stout vertical plates, which curve downward below and unite to meet the ilia. Each vertebra supports its own process, although there is a tendency to overlap in front. There is a gradual increase in size from the first to the last sacral vertebra, and the first caudal is larger than the last sacral. The neural cavity of the sacrum is described below.

THE CAUDAL VERTEBRÆ.

The caudal vertebræ present the greatest diversity, both in size and form. The anterior caudals are the largest in the whole vertebral series, and are highly modified to support a portion of the massive dermal armor. The articular faces of their centra are nearly plane and very rugose. The neural spine has an enormous development, and its summit is expanded into a bifurcate rugose head. These caudals are very short, and their neural spines nearly or quite in apposition above. Their centra have no distinct faces for chevrons. The transverse processes are expanded vertically, and their extremities curve downward. Farther back the same general characters are retained, but the centra are more deeply cupped and the spines less massive. Figs. 5, 6, and 7 of Pl. XLV show anterior caudal vertebræ. The chevrons here have their articular ends separate and rest upon two vertebræ, as shown on the same plate.

In the median caudals, figs. 8 and 9, the spine has greatly diminished in height, and the faces for chevrons are placed on prominent tubercles on the postero-inferior surface. The lower margin of the front articular face is sharp and the chevrons do not meet it. In the more distal caudals the neural spine and zygapophyses are reduced to mere remnants, but the chevron facets remain distinct. These vertebræ, as well as those farther back, have their centra much compressed. The caudal vertebræ are remarkably uniform in length throughout most of the series.

THE SACRAL CAVITY.

During an investigation of one skeleton of this genus the writer found a very large chamber in the sacrum, formed by an enlargement of the spinal canal. This chamber was ovate in form, and strongly resembled the brain case in the skull, although very much larger, being at least twenty times the size of the cavity which contained the brain. This remarkable feature led to the examination of the sacra of several other individuals of Stegosaurus, and it was found that all had a similar large chamber in the same position. The form and proportions of this cavity are indicated in Pl. XLVI, figs. 2 and 3, which represent a cast of the entire neural canal inclosed in the sacrum. The large vaulted chamber, it will be observed, is contained mainly in the first and second sacral vertebræ, although the canal is considerably enlarged behind this cavity. The sections represented in fig. 4 are in each case made where the transverse diameters are greatest.

The remarkable feature about this posterior brain case, if so it may be called, is its size in comparison with that of the true brain of the animal, which is also indicated in the same plate, and in this respect it is entirely without a parallel. A perceptible swelling in the spinal cord of various recent animals has indeed been observed in the pectoral and pelvic regions, where the nerves are given off for the anterior and posterior limbs; and in extinct forms some very noticeable cases are recorded, especially in dinosaurs, but nothing that approaches the sacral enlargement in Stegosaurus has hitherto been known. The explanation may doubtless in part be found in the great development of the posterior limbs in this genus; but in some allied forms—Camptosaurus, for example, where the disproportion between the fore and hind limbs is quite as marked—the sacral enlargement of the spinal cord is not one-fourth as great as in Stegosaurus.

It is an interesting fact that in young individuals of Stegosaurus the sacral cavity is proportionately larger than in adults, which corresponds to a well-known law of brain growth.

The physiological effect of a posterior nervous center so many times larger than the brain itself is a suggestive subject which need not here be discussed. It is evident, however, that in an animal so endowed the posterior part was dominant.

THE PELVIS.

The ilium in Stegosaurus is a very peculiar bone, unlike any hitherto known in the reptiles. Its most prominent feature is its great anterior extension in front of the acetabulum. Another striking character is seen in its superior crest, which curves inward, and firmly unites with the neural arches of the sacrum, thus roofing over the cavities between the transverse processes. The acetabular portion of the ilium is large and shallow (Pl. XLVI, fig. 5). The face for union with the ischium is large and rugose, but that for the pubis is much less distinct. The postacetabular part of the ilium is very short, scarcely one-third as long as the anterior projection.

The ischium of *Stegosaurus ungulatus* is short and robust, and has a prominent elevation on the upper margin of the shaft (Pl. XLVIII, fig. 2). Its larger articular face meets a postacetabular process of the ilium, and a smaller articulation joins the pubis. The shaft of the ischium is twisted so that it resembles somewhat the corresponding bone of Morosaurus. The pelvis of *Stegosaurus stenops* is shown in the same plate, fig. 3.

The pubic element of the pelvis of *Stegosaurus ungulatus* is in general form somewhat like that of Camptosaurus. The prepubis consists of a strong spatulate process, projecting forward nearly horizontally. Its proximal end articulates with the preacetabular process of the ilium. The postpubic branch extends backward and downward, nearly to the end of the ischium. The two bones fit closely together in this region. The usual pubic foramen is in this species replaced by a notch, opening into the acetabular cavity. In a smaller species, *Stegosaurus affinis*, the postpubic bone is slender and more rod-like, not flattened as in the specimen here figured.

THE FORE LIMBS.

On Pl. XLVII some of the bones of the scapular arch and anterior limbs of Stegosaurus are figured. The scapula and coracoid are of the true dinosaurian type. The former has its upper portion rather short and moderately expanded. The coracoid was closely united to the scapula by cartilage. It is perforated by the usual foramen, which in some cases may become a notch.

The humerus (fig. 3) is short and massive. It has a distinct head and a strong radial crest. The shaft is constricted medially, and is without any medullary cavity. The ulna (fig. 4) is also massive, and has a very large olecranal process. Its distal end is comparatively small. The radius is smaller than the ulna. The fore limb, as a whole, was very powerful and adapted to varied movements. There were five well-developed digits in the fore foot, as shown on Pl. XLVII, fig. 1.

THE HIND LIMBS.

The femur of Stegosaurus (Pl. XLVII, fig. 1) is by far the largest bone in the skeleton. It is remarkably long and slender. There is no distinct head, and the great trochanter is nearly or quite obsolete. The shaft is of nearly uniform width and very straight. There is no evidence of a third trochanter. The distal end of the femur is peculiar in having very flat condyles, with only a shallow depression between them. The external one has only a rudiment of the ridge which passes between the heads of the tibia and fibula, and is very characteristic of true dinosaurs and birds.

The tibia (fig. 2) is very much shorter than the femur. Its superior end is unusually flat, indicating that it met the condyles of the femur so as to bring the two bones at times nearly or quite into the same line. The shaft of the tibia is constricted medially, leaving a wide space between it and the fibula. The distal end of the tibia is blended entirely with the convex astragalus, so as to resemble strongly the corresponding part in birds.

The fibula (fig. 2) is slender, and has its smaller end above. This extremity is applied closely to the head of the tibia by a rugose suture, so as to unite readily with it. Its upper articular surface is nearly or quite on a level with that of the tibia. The distal end of the fibula is

expanded, and in the specimen figured is firmly coossified with the calcaneum. The two coalesce with the tibia and astragalus, and form a
smooth convex articulation for the ankle joint. There were only three
functional digits in the hind feet, as shown on Pl. XLVIII, fig. 2.

THE DERMAL ARMOR.

The most remarkable feature about Stegosaurus is the series of ossifications which formed its offensive and defensive armor. These consist
of numerous spines, some of great size and power, and many bony
plates of various sizes and shapes, well fitted for protecting the animal
against assaults. Some of these plates are a meter, or more than 3 feet,
in diameter.

The spines were of different forms and varied much in size. Four
of these are represented on Pl. L. All of those preserved are unsym-
metrical, and most of them are in pairs. One of the largest is shown
in fig. 2, which gives the more usual form and proportions. This specimen is over 2 feet in length.

The osseous dermal covering of the Stegosauria was first described
by the writer from specimens found associated with several skeletons,
but not in place, and hence the position of the various parts was a
matter of considerable doubt. Subsequent discoveries have shown the
general arrangement of the plates, spines, and ossicles. and it is now
evident that, while all the group were apparently well protected by
offensive and defensive armor, the various species, and perhaps the
sexes, differed more or less in the form, size, and number of portions of
their dermal covering. This was especially true of the spines, which
are quite characteristic in some members of the group, if not in all.

The skull was evidently covered above with a comparatively soft
integument. The throat and neck below were well protected by small,
rounded and flattened ossicles having a regular arrangement in the
thick skin. One of these ossicles is shown in Pl. XLIX, fig. 1. The
upper portion of the neck, back of the skull, was protected by plates,
arranged in pairs on either side. These plates increased in size farther
back, and thus the trunk was shielded from injury. From the pelvic
region backward a series of huge plates stood upright along the median
line, gradually diminishing in size to about the middle of the tail. One
of these is shown in Pl. XLIX, fig. 4. Some of the species, at least, had
somewhat similar plates below the base of the tail, and one of these
bones is represented in fig. 3 of the same plate.

The offensive weapons of this group were a series of huge spines
arranged in pairs along the top of the distal portion of the tail, which
was elongate and flexible, thus giving effective service to the spines,
as in the genus Myliobatis.

In *Stegosaurus ungulatus* there were four pairs of these spines,
diminishing in size backward. Two of the larger of these are shown
on Pl. L, figs. 2 and 3. In some other forms there were three pairs,
and in *S. stenops* but two pairs have been found.

In one large species, *Stegosaurus sulcatus*, there is at present evidence of only one pair of spines. These are the most massive of any yet found, and have two deep grooves on the inner face, which distinguish them at once from all others known. One of these grooved spines is represented on Pl. L, figs. 4, 5, and 6.

The position of these caudal spines with reference to the tail is indicated in the specimen figured on Pl. LI, which shows the vertebræ, spines, and plate as found.

DIRACODON.

The American genera of the Stegosauria are Stegosaurus and Diracodon. Of the former there are several well-marked species besides *S armatus*, the type. Of the latter genus but one is known at present, *Diracodon laticeps*, the remains of which have hitherto been found at a single locality only, where several individuals referred to this species have been discovered. Aside from the form of the skull, these specimens have in the fore foot the intermedian and ulnar bones separate, while in Stegosaurus these carpals are firmly coossified.

All the American Stegosauridæ have the second row of carpals unossified and five digits in the manus. In the hind foot the astragalus is always coossified with the tibia, even in very young specimens, while the calcaneum is sometimes free. The second row of tarsals is not ossified in any of the known specimens. Only four digits in the hind foot are known with certainty, and one of these is quite small. All forms have at least three well-developed metatarsals, which are short and massive, but longer and much larger than the metacarpals.

In one large specimen, of which the posterior half of the skeleton was secured, no trace of dermal armor of any kind was found. If present during life, as indicated by the massive spines of the vertebræ, it is difficult to account for its absence when the remains were found, unless, indeed, the dermal covering had been removed after the death of the animal and previous to the entombment of the skeleton where found. In this animal the ilia were firmly coossified with the sacrum, thus forming a strong bony roof over the pelvic region, as in birds.

This specimen represents a distinct species, *Stegosaurus duplex*. It was originally referred by the writer to *S. ungulatus*. In the sacrum of this species each vertebra supports its own transverse process, or rib, as in the Sauropoda, while in *S. ungulatus* the sacral ribs have shifted somewhat forward, so that they touch, also, the vertebra in front, thus showing an approach to some of the Ornithopoda.

CHARACTERS OF STEGOSAURIA.

The large number of specimens of the Stegosauria now known from the American Jurassic, and the fine preservation of some of the remains, aid in forming a more accurate estimate of the relations of the group to the other dinosaurs than has hitherto been possible. The presence of a predentary bone, and the well developed postpubis, are important characters that point to the Ornithopoda as near allies, with a common ancestry. These positive characters are supplemented by some points in the structure of the skull and the form of the teeth.

There are, however, a large number of characters in which the Stegosauria differ from the Ornithopoda, and among these are the following:

(1) All the bones of the skeleton are solid.
(2) The vertebræ are all biconcave.
(3) All the known forms have a strong dermal armor.
(4) The second row of carpals and tarsals are unossified.
(5) The astragalus is coossified with the tibia.
(6) The spinal cord was greatly enlarged in the sacral region.

RESTORATION OF STEGOSAURUS.

PLATE LII.

In this restoration the animal is represented as walking, and the position is adapted to that motion. The head and neck, the massive fore limbs, and, in fact, the whole skeleton, indicate slow locomotion on all four feet. The longer hind limbs and the powerful tail show, however, that the animal could thus support itself, as on a tripod, and this position could perhaps have been easily assumed in consequence of the massive hind quarters.

In the restoration as here presented the dermal armor is the most striking feature, but the skeleton is almost as remarkable, and its high specialization was evidently acquired gradually as the armor itself was developed. Without the latter many points in the skeleton would be inexplicable, and there are still a number that need explanation.

The small, elongated head was covered in front by a horny beak. The teeth are confined to the maxillary and dentary bones, and are not visible in the figure here given. They are quite small, with compressed, fluted crowns, and indicate that the food of this animal was soft, succulent vegetation. The vertebræ are solid, and the articular faces of the centra are biconcave or nearly flat. The ribs of the trunk are massive and placed high above the centra, only the tubercle being supported on the elevated diapophysis. The neural spines, especially those of the sacrum and anterior caudals, have their summits expanded to aid in supporting the massive dermal armor above them. The limb bones are solid, and this is true of every other part of the skeleton. The feet were short and massive, and the terminal phalanges of the functional toes were covered by strong hoofs. There were five well-developed digits in the fore foot, and only three in the hind foot, the first toe being rudimentary and the fifth entirely wanting.

In life the animal was protected by a powerful dermal armor, which served both for defense and offense. The throat was covered by a thick skin, in which were embedded a large number of rounded ossicles, as shown in the plate. The gular portion represented was found beneath

the skull, so that its position in life may be regarded as definitely settled. The series of vertical plates which extended above the neck,
along the back, and over two-thirds of the tail. is a most remarkable
feature, which could not have been anticipated and would hardly have
been credited had not the plates themselves been found in position.
The four pairs of massive spines characteristic of the present species,
which were situated above the lower third of the tail, are apparently
the only part of this peculiar armor used for offense. In addition to
the portions of armor above mentioned there was a pair of small plates
just behind the skull, which served to protect this part of the neck.
There were also, in the present species, four flat spines, which were
probably in place below the tail, but as their position is somewhat in
doubt they are not represented in the present restoration.

All these plates and spines, massive and powerful as they now are,
were in life protected by a thick, horny covering, which must have
greatly increased their size and weight. This covering is clearly indicated by the vascular grooves and impressions which mark the surface
of both plates and spines, except their bases, which were evidently
implanted in the thick skin.

The peculiar group of extinct reptiles named by the writer the Stegosauria, of which a typical example is represented in the present restoration, is now nearly as well known as any other dinosaurs. They are
evidently a highly specialized suborder of the Predentata, which have
the Ornithopoda as their most characteristic members, and all doubtless had a common ancestry.

Another highly specialized branch of the same great order is seen
in the gigantic Ceratopsia of the Cretaceous, which the writer has
recently investigated and made known. The skeleton of the latter
group presents many interesting points of resemblance to that of
the Stegosauria, which can hardly be the result of adaptation alone,
but the wide difference in the skull and in some parts of the skeleton
indicates that their affinities are remote. A comparison of the present
restoration with that of Triceratops on Pl. LXXI will make the contrast between the two forms clearly evident.

DISTRIBUTION OF STEGOSAURIA.

All the typical members of the Stegosauria are from the Jurassic
formation, and the type specimen used in the present restoration was
found in Wyoming, in the Atlantosaurus beds of the upper Jurassic.
Diracodon, a genus nearly allied to Stegosaurus, occurs in the same
horizon. Palæoscincus Leidy, 1856, from the Cretaceous, and Priconodon of the writer, 1888, from the Potomac formation, are perhaps allied
forms of the Stegosauria, but until additional remains are found their
exact affinities can not be determined. Apparently the oldest known
member of this group in America is the Dystrophæus Cope, 1877, from
the Triassic of Utah.

One of the best-preserved specimens of the Stegosauria in Europe was described by Owen, in 1875, as *Omosaurus armatus*, and the type specimen is in the British Museum. It is from the Kimmeridge clay (Upper Jurassic) of Swindon, England The skull is wanting, but the more important parts of the skeleton are preserved. Various portions of the skeleton of several other individuals have also been found in England, but the skull and teeth still remain unknown.

Another genus of the Stegosauria, representing a distinct family, is Scelidosaurus, established by Owen in 1859, from the Lias of England. The greater part of the skeleton is known. A restoration in outline, made by the writer for comparison with that of Stegosaurus, is shown on Pl. LXXXIII. The Euskelesaurus of Huxley, 1867, from the Trias of South Africa, is apparently a member of this group.

CAMPTOSAURIDÆ.

Another important family of Jurassic Dinosaurs is the Camptosauridæ, so named from the type genus Camptosaurus, described by the writer in 1879, the type specimen being from the Atlantosaurus beds of Wyoming. This genus includes several species of herbivorous dinosaurs, which belong to the true Ornithopoda, or bird-footed forms. The species were all bipedal, with the fore limbs much smaller than those behind, and all the limb bones light and hollow. The head was comparatively small, the neck of moderate length and quite flexible, and in life the animals were evidently agile and graceful in movement. Some of the smaller species were quite bird-like in form and structure. The three American genera, Camptosaurus, Dryosaurus, and Laosaurus, are all from the same general horizon.

CAMPTOSAURUS.

The large dinosaur described by the writer as *Camptosaurus dispar*, of which a restoration is given on Pl. LVI, is now so well known that it may be taken as a form typical of the group. It is exceeded in size by *Camptosaurus amplus* Marsh, but there are at least two smaller species of the genus (*C. medius* and *C. nanus*). So far as at present known these species are found in successive deposits of the same general horizon, the smallest below and the largest above.

Camptosaurus amplus is represented by remains which show that this reptile when alive was about 30 feet in length. The type specimen of *C. dispar* was about 20 feet in length and 10 feet in height. *C. medius* was about 15 feet long. The smallest species of the genus, *C. nanus*, was not more than 10 feet in length, and perhaps 6 feet in height when standing at rest. One of the striking features of this diminutive species is its long sigmoid scapula, shown in fig. 2, Pl. LV. This is in strong contrast with the short, straight scapula of *C. dispar*, seen on Pl. LIV, fig. 1. The limb bones of all the species of this genus are very hollow.

The skull, brain, and teeth of *C. medius* are shown on Pl. LIII. The peculiar peg-and-notch articulation in the sacral vertebræ of this genus, already described elsewhere, is indicated on Pl. LIV, figs. 3 and 4, and a summary of the principal characters of the genus, and of the nearest allied genera, will be found on p. 201.

RESTORATION OF CAMPTOSAURUS.

PLATE LVI.

The restoration here given is based upon the type specimen of *Camptosaurus dispar*, one of the most characteristic forms of the great group Ornithopoda, or bird-footed dinosaurs. The reptile is represented on Pl. LVI, one-thirtieth natural size. The position chosen was determined after a careful study not only of the type specimen, but of several others in excellent preservation, belonging to the same species or to others nearly allied. It is therefore believed to be a position frequently assumed by the animal during life, and thus, in some measure, characteristic of the genus Camptosaurus. The type specimen of the present species, when alive, was about 20 feet in length, and 10 feet high in the position here represented.

The genus Camptosaurus is a near ally of Iguanodon of Europe, and may be considered its American representative. Camptosaurus, however, is a more generalized type, as might be expected from its lower geological horizon. It resembles more nearly some of the Jurassic forms in England generally referred to Iguanodon, but as these are known only from fragmentary specimens their generic relations with Camptosaurus can not now be determined with certainty.

In comparing Camptosaurus, as here restored, with a very perfect skeleton of Iguanodon from Belgium, as described and figured, various points of difference as well as of resemblance may be noticed. The skull of Camptosaurus had a sharp, pointed beak, evidently encased during life in a horny sheath. This was met below by a similar covering, which inclosed the predentary bone. The entire front of the upper and lower jaws was thus edentulous, as in Iguanodon, but of different shape. The teeth of the two genera are of similar form, and were implanted in like manner in the maxillary and dentary bones. In Camptosaurus there is over each orbit a single supraorbital bone, curving outward and backward, with a free extremity, as in the existing monitor, a feature not before observed in any other dinosaur except Laosaurus, an allied genus, also from the Jurassic of America. Other portions of the skull of Camptosaurus, as well as the hyoid bones, appear to agree in general with those of Iguanodon.

The vertebræ of Camptosaurus are similar in many respects to those of Iguanodon, but differ in some important features. In the posterior dorsal region the transverse processes support both the head and tubercle of the rib, the head resting on a step, as in existing crocodiles. The five sacral vertebræ, moreover, are not coossified even in adult forms, and to this character the name Camptonotus, first given to the genus by the writer in 1879, especially refers. Another notable feature of the sacral vertebrae of the type specimen should be mentioned. The vertebrae of the sacrum, especially the posterior four, are joined to each other by a peculiar peg-and-notch articulation. The floor of the neural canal of each vertebra is extended forward into a pointed process (somewhat like an odontoid process), which fits into a corresponding cavity of the centrum in front. This arrangement, while permitting some motion between the individual vertebrae, helps to hold them in place, thus compensating in a measure for absence of anchylosis. A similar method of articulation is seen in the dermal scales of some ganoid fishes, but so far as the writer is aware nothing of the kind has been observed before in the union of vertebrae.

In Camptosaurus the sternum was entirely unossified, and no trace of clavicles has been found. The pelvis of Camptosaurus differs especially from that of Iguanodon in the pubis, the postpubic branch being even longer than the ischium, while in Iguanodon this element is much shortened.

In the fore foot of Camptosaurus there were five functional digits, the first being flexible and nearly parallel with the second, thus differing from the divergent, stiff thumb of Iguanodon. The hind feet had each three functional digits only, the first being rudimentary and the fifth entirely wanting, as shown in Pl. LVI. The entire skeleton of Camptosaurus was proportionately more slender and delicately formed than that of Iguanodon, although the habits and mode of life of these two herbivorous dinosaurs were doubtless very similar.

The type specimen of *Camptosaurus dispar*, used as the basis of the present restoration, is from the Atlantosaurus beds of the Upper Jurassic of Wyoming. This species and other allied forms will be described in full in an illustrated memoir now in preparation by the writer for the United States Geological Survey. The present restoration is reduced from a large drawing made for that volume.

LAOSAURIDÆ.

DRYOSAURUS.

Another genus of Jurassic dinosaurs, allied to Camptosaurus, but differing from it in many important respects, is Dryosaurus. The type was described by the writer in 1878 under the name *Laosaurus altus*, and a tooth, the pelvis, and a hind leg were also figured. Additional material since received shows that this genus is quite distinct from Laosaurus, to which it was at first referred, and is intermediate between Camptosaurus and that genus, as is shown in a summary of the characters of these genera given later in the present article.

The only species of Dryosaurus at present known is the type first described, and this form is now called *Dryosaurus altus* (Pl. LV, fig. 4). Several specimens of this dinosaur are preserved in the Yale

museum, and they show it to have been in life about 10 or 12 feet
long, and one of the most slender and graceful members of the group.
The known remains are all from the Atlantosaurus beds of Colorado
and Wyoming.

LAOSAURUS.

The present genus includes several species of diminutive dinosaurs,
all much smaller than those above described, and possessing many features now seen only in existing birds, especially in those of the ostrich
family. The two species of the genus first described by the writer
(*Laosaurus celer*, the type, and *Laosaurus gracilis*) show these avian
features best of all, and it would be difficult to tell many of the isolated
remains from those of birds. A larger species, which has been called
Laosaurus consors, is known by several skeletons nearly complete.
The type specimen, here figured in part on Pl. LV, figs. 1 and 3, is the
most perfect of all, and this was collected by the writer in 1879. The
animal when alive was about 8 or 10 feet in length. The known remains
are from the Atlantosaurus beds of Wyoming.

One of the distinctive features of this genus, which separates it at
once from those above described, is the pubis. The prepubis, or anterior branch of this bone, which was very large and broad in Camptosaurus, still long and spatulate in Dryosaurus, is here reduced to a
pointed process not much larger than in some birds. These differences
are shown in Pl. LIV and in Pl. LV, figs. 3 and 4.

The European representative of Laosaurus is Hypsilophodon Huxley,
from the Wealden of England. That genus, however, differs from the
nearest allied forms of this country in several well-marked characters.
Among these the presence of teeth in the premaxillary bones and a
well-ossified sternum are features not seen in American Jurassic forms.
The fifth digit of the manus, moreover, in Hypsilophodon is almost at
right angles to the others, and not nearly parallel with them, as in
Dryosaurus. It agrees with the latter genus in having the tibia longer
than the femur. An outline restoration of Hypsilophodon, made by
the writer for comparison with allied American forms, is given on Pl.
LXXXIV.

NANOSAURIDÆ.

NANOSAURUS.

The smallest known dinosaur, representing the type species of the
present genus, was described by the writer in 1877, under the name
Nanosaurus agilis. The type specimen consists of the greater portion
of the skull and skeleton of one individual, with the bones more or less
displaced and all entombed in a slab of very hard quartzite. The whole
skeleton was probably thus preserved in place, but before its discovery
a part of the slab had been split off and lost. The remaining portion
shows on the split surface many important parts of the skeleton, and
these have been further exposed by cutting away the matrix, so that

the main characters of the animal can be determined with considerable certainty.

A study of these remains shows that the reptile they represent was one of the typical Ornithopoda, and one of the most bird-like yet discovered. A dentary bone in fair preservation (fig. 42) indicates that the animal was herbivorous, and the single row of pointed and compressed teeth, thirteen in number and small in size, forms a more regular and uniform series than in any other member of the group. The ilium, also, shown in fig. 43, is characteristic of the Ornithopoda, having a slender, pointed process in front, but one much shorter than in any of

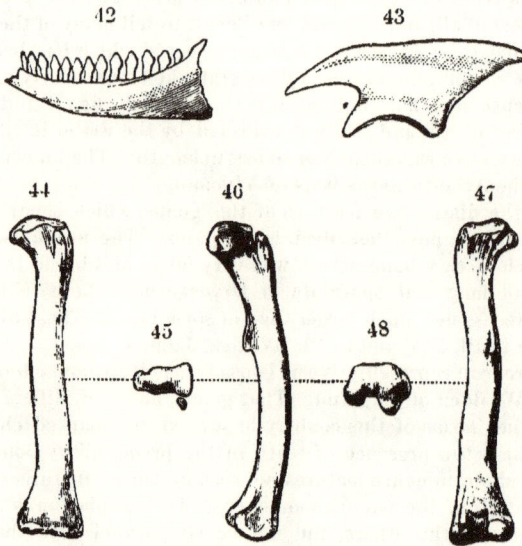

Fig. 42.—Dentary bone of *Nanosaurus agilis* Marsh; seen from the left.
Fig. 43.—Ilium of same individual; left side.
Both figures are natural size.
Fig. 44.—Left femur of *Nanosaurus rex* Marsh; front view.
Fig. 45.—Proximal end of same.
Fig. 46.—The same bone; side view.
Fig. 47.—The same; back view.
Fig. 48.—Distal end of same.
All five figures are one-half natural size.

the larger forms. The posterior end is also of moderate size. All the bones of the limbs and feet are extremely hollow, strongly resembling in this respect those of birds. The femur was shorter than the tibia. The metatarsals are greatly elongated and very slender, and there were probably but three functional toes in the hind foot.

A second form referred by the writer to this genus, under the name *Nanosaurus rex*, may perhaps belong to the genus Laosaurus. The femur is shown in figs. 44 to 48, above. The animal thus represented was considerably larger than the present type species and from a some-what higher horizon in the Atlantosaurus beds.

The type specimen here described, which pertained to an animal about half as large as a domestic fowl, was found in Colorado. This reptile was a contemporary of the carnivorous Hallopus, likewise one of the most diminutive of dinosaurs, and one of the most remarkable.

DETERMINATION OF GENERA.

The various dinosaurs thus briefly referred to under their respective genera have many other points of interest that can not be here dis-cussed, but their resemblance to birds is worthy of some notice. This is apparent in all of them, but in the diminutive forms the similarity becomes more striking. In all the latter the tibia is longer than the femur, a strong avian character, and one seen in dinosaurs only in the small bird-like forms.[1] In Nanosaurus nearly all, if not all, the bones preserved might have pertained to a bird, and the teeth are no evidence against this idea. In the absence of feathers an anatomist could hardly state positively whether this was a bird-like reptile or a reptilian bird.

The main characters of the four genera above discussed are as follows:

CAMPTOSAURUS.

Premaxillaries edentulous, with horny beak. Teeth large, irregular, and few in number. A supraorbital fossa. Cervical vertebræ long and opisthocœlous. Lumbars present. Five vertebræ in sacrum, with peg-and-notch articulation. Sternum unossified. Limb bones hollow. Fore limbs small. Five functional digits in manus. Prepubis long and broad ; postpubis elongated. Femur longer than tibia. Metatar-sals short. Three functional digits in pes, the first rudimentary and the fifth wanting.

DRYOSAURUS.

Premaxillaries edentulous, with horny beak. Teeth of moderate size. A supraorbital fossa. Cervicals long and biconcave. No lumbars. Six coossified vertebræ in sacrum, without peg-and-notch articulation. Sternum unossified. Limb bones hollow. Fore limbs very small. Five digits in manus. Prepubis long and narrow; postpubis elongate and slender. Posterior limbs very long. Femur shorter than tibia. Meta-tarsals long and hollow. First digit in pes complete ; fifth metatarsal represented by short splint only.

LAOSAURUS.

Premaxillaries edentulous. Teeth small and irregular. Cervicals short and flat. Six coossified vertebræ in sacrum; no peg-and-notch articulation. Sternum unossified. Fore limbs small. Limb and foot bones hollow. Prepubis very short and pointed; postpubis slender. Femur shorter than tibia. Metatarsals elongate. First digit in pes functional; fifth rudimentary.

---

[1] Besides the genera here mentioned, Cœlurus, Compsognathus, and Hallopus also possess this character.

NANOSAURUS.

Teeth compressed and pointed, and in a single uniform row. Cervical and dorsal vertebræ short and biconcave. Sacral vertebræ three (?). Anterior caudals short. Ilium with very short, pointed front and narrow posterior end. Fore limbs of moderate size. Limb bones and others very hollow. Femur curved and shorter than tibia. Fibula pointed below. Metatarsals very long and slender.

The genera thus defined contain all the known forms of the typical Ornithopoda from the American Jurassic. They are, moreover, the earliest representatives of this group known in this country from osseous remains, as such fossils have not yet been found in the Triassic, where the oldest dinosaurs occur. Some of the bird-like footprints in the Connecticut River sandstone may indeed have been made by dinosaurs of this group, but there is no positive evidence on this point. The American Cretaceous forms of the typical Ornithopoda, so far as at present known, are all of large size and highly specialized, and this appears to be true also of the Old World species.

RESTORATION OF LAOSAURUS.

PLATE LVII.

The present restoration in outline of *Laosaurus consors*, one-tenth natural size, will serve to show the form and proportions of one of the most bird-like of the smaller Jurassic Ornithopoda and its contrast with the more massive Camptosaurus from the same horizon. A comparison of this restoration with that of Hypsilophodon from the English Wealden (Pl. LXXXIV) is especially instructive, as the two animals were near allies, although from different geological horizons.

The position here chosen for the restoration of Laosaurus is one which would seem to have been natural to the animal when standing at rest. This would mean a height of about 4 feet, with 8 feet in length. That the animal was bipedal in its usual locomotion on land is assumed in this case from the general structure, especially the very small and weak fore limbs, and the large size and strong articulations of the posterior limbs. When walking upright, as here represented, it seems probable that the animal would touch the ground with its tail; but this is by no means certain. That reptiles of similar structure and proportions could walk on their hind feet without leaving a mark of the tail is clearly indicated by many long series of bipedal footprints left in the sandstone of the Connecticut Valley, some of which have already been described and figured in this paper. In the present species the tail was powerful and more or less compressed, thus suggesting its use in swimming.

PART III.

CRETACEOUS DINOSAURS.

During Cretaceous time in North America the dinosaurs were still abundant, and most of them were much more specialized than those that lived in the preceding periods. Some of the Cretaceous forms were the strangest of the whole group, of gigantic size and bizarre appearance. Others were diminutive in size and so bird-like in form and structure that their remains can be distinguished with difficulty, if at all, from those of birds.

Of the carnivorous dinosaurs known from Cretaceous deposits the family Dryptosauridæ is especially conspicuous, on account of the large size and ferocious nature of all its representatives. In the later Cretaceous a second family, the Ornithomimidæ, was also abundant, and among its members were some of the most minute and bird-like of dinosaurs hitherto discovered. Of the herbivorous forms the huge, horned Ceratopsidæ, the most remarkable of all dinosaurs, were for a limited period the dominant reptiles in western North America. Living at the same time with these were the Claosauridæ, large bipedal dinosaurs, of sluggish disposition, that dwelt along the shores of the lakes and rivers of that time. Besides these were still others related to the Jurassic Stegosaurus, among them the Nodosauridæ, quadrupedal forms with heavy dermal armor. All these became extinct at the close of the Cretaceous, and no remains of dinosaurs have been found in place in any later deposits.

THEROPODA.

DRYPTOSAURIDÆ.

This family is well represented throughout the Cretaceous in North America, but up to the present time only fragmentary remains have been found, so that little is known about the skull, pelvis, and feet, the most characteristic portions of the skeleton. So far as now determined they appear to be nearly allied to Megalosaurus of Europe and include Allosaurus from the Jurassic of this country.

Remains of the genus Dryptosaurus (Lælaps) have been found at various localities on the Atlantic Coast, especially in the marl region of New Jersey. Many of these fossils have been described by Prof. Cope.[1]

ORNITHOMIMIDÆ.

This family can be separated sharply from all other dinosaurs by the hind feet, which contain three functional metatarsals, the middle one, or third, of which has its proximal end much diminished in size and crowded backward behind the second and fourth, as in many existing birds.

---

[1] Extinct Batrachia, Reptilia, and Aves of North America, p. 100, 1870.

Another important character, seen also in the genus Ceratosaurus of the Jurassic, is found in the pelvis, the bones of which are firmly coossified with each other and with the sacrum.

ORNITHOMIMUS.

The most marked characters of the genus Ornithomimus already determined are manifest in the limbs and feet, and these have been selected for description in the present article. A typical example is shown on Pl. LVIII, figs. 1–4, which is the type specimen, the species being *Ornithomimus velox* Marsh.

On the distal part of the tibia represented in fig. 1 the astragalus is seen in place, with a very large ascending process, larger than in any dinosaur hitherto known. The calcaneum is also shown in position, but the slender fibula is absent. This bone was complete, but of little functional value. The tibia and all the larger limb bones were hollow, with thin walls, as indicated in the section, fig. 1, *c*.

In fig. 5 the corresponding parts of a young ostrich are shown for comparison. The slender, incomplete fibula is in place beside the tibia. The astragalus with its ascending process, and the distinct calcaneum, are also shown in position. The almost exact correspondence of these different parts in the bird and reptile will be manifest to every anatomist.

THE METATARSALS.

The most striking feature of the foot belonging with the reptilian tibia is shown in the metatarsals represented in fig. 2, *A*. These are three in number, and are in the same position as in life. They are the three functional metatarsals of the typical Ornithopoda and of birds. The distal ends of these bones correspond in size and relative position in the two groups, but here, in the present specimen, the reptilian features cease, and those of typical birds replace them. In all the reptiles known hitherto, and especially in dinosaurs, the second, third, and fourth metatarsals are prominent in front, at their proximal ends, and the third is usually the largest and strongest. In birds the place of the third is taken above by the second and fourth, the third being crowded backward and very much diminished in size.

This character is well shown in fig. 6, which represents the second, third, and fourth metatarsals of a young turkey, with the tarsal bones absent. In the reptilian metatarsals seen in fig. 2 the same arrangement is shown, with the tarsals in place. The second and fourth metatarsals have increased much in size in the upper portion, and meet each other in front.

The third metatarsal, usually the largest and the most robust throughout, here diminishes in size upward, and takes a subordinate, posterior position, as in birds. The correspondence between the metatarsals of the bird and reptile are here as strongly marked as in the tibiae and their accompanying elements, above described.

In fig. 3 of the plate the three phalanges represented belong with
the second metatarsal, and were found together in place.

The three metacarpals represented in fig. 4 were found together in
position, near the remains of the hind limb here described. Their very
small size is remarkable, and they may possibly belong to a smaller
individual, but with this exception there is no reason why they do not
pertain to the same specimen as the hind foot. The remains of this
species were found by George L. Cannon, jr., in the Ceratops beds of
Colorado.

THE PELVIC ARCH.

A larger species from the same horizon, *Ornithomimus sedens*, more
recently described by the writer, is based upon the nearly complete

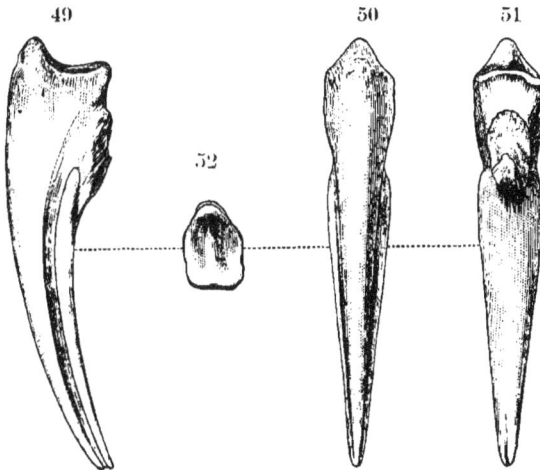

Fig. 49.—Terminal phalanx, manus of *Ornithomimus sedens* Marsh; side view.
Fig. 50.—The same phalanx; front view.
Fig. 51.—The same; back view.
Fig. 52.—Proximal end of same.
All the figures are one-half natural size.

pelvis, with various vertebræ, and some other parts of the skeleton.
The most striking feature of the pelvis is the fact that the ilium,
ischium, and pubis are firmly coossified with one another, as in recent
birds. This character has been observed hitherto among dinosaurs
only in the genus Ceratosaurus, described by the writer from the Jurassic of Wyoming. The present pelvis resembles that of Ceratosaurus
in its general features, but there is no foramen in the pubis.

There are five vertebræ in the sacrum, firmly coossified with one
another, as are also the sacral spines. The sacral vertebræ are grooved
below, with the sides of the centra excavated. The caudals have the
diplosphenal articulation, and the first caudal bears a chevron. All
the bones preserved are very delicate, and some of them, at least, are

apparently pneumatic. The sacrum measures 15 inches in length, and the twelve caudals following occupy a space of 31 inches. The known remains indicate a reptile about 8 or 10 feet in length. A terminal phalanx of the fore foot is represented on page 205.

In the same horizon occur the remains of a very minute species, which agrees in all its characters, so far as determined, with the members of this genus. The most characteristic portions secured are the metatarsal bones, and these show the same features exhibited in the type species of the genus, *O. relox.* They are, however, so much smaller as to suggest that they may pertain to a bird. Various portions of the second, third, and fourth metatarsals are known, and the distinctive feature is seen in the third, which has the upper part of the shaft so attenuated that it may not reach to the tarsus. The second and fourth metatarsals are very long and slender. This unique animal was about the size of the common fowl. The species has been called *Ornithomimus minutus.*

The large species described by the writer as *Ornithomimus grandis* belongs in essentially the same horizon. Portions of two other skeletons have since been obtained, which apparently pertain to this species. In one of these the femur, tibia, and fibula are in good preservation, and they clearly demonstrate that this reptile was one of the largest of the Theropoda. The femur and tibia have each a very large cavity in the shaft, with well-defined walls. Even the fibula has a cavity in its upper portion. In the other specimen the second metatarsal is in fair preservation and shows the same form as in the type of the genus.

There is much probability that this gigantic carnivore was one of the most destructive enemies of the herbivorous Ceratopsidæ, next to be described.

PREDENTATA.

CERATOPSIDÆ.

The huge horned dinosaurs, from the Cretaceous, recently described by the writer,[1] have now been investigated very carefully, and much additional light has been thrown upon their structure and affinities. A large amount of new material has been secured, including several skulls, nearly complete, as well as various portions of the skeleton.

CERATOPS BEDS.

The geological deposits, also, in which their remains are found have been carefully explored during the last few years, and the known localities of importance examined by the writer, to ascertain what other fossils occur in them and what were the special conditions which preserved so many relics of this unique fauna. The definite horizon in which these strange reptiles occur has been called by the writer the Ceratops beds, from the type genus Ceratops, and its position is shown in the section on page 145.

---

[1] American Journal of Science, 1888-1894.

This geological horizon is a distinct one in the upper Cretaceous, and is indicated for more than 800 miles along the eastern flank of the Rocky Mountains. It is marked at nearly every outcrop by remains of these reptiles, and hence the strata containing them have been called the Ceratops beds. They are fresh-water or brackish deposits which form a part of the so-called Laramie, but are below the uppermost beds referred to that group. In some places, at least, they rest upon marine beds, which contain invertebrate fossils characteristic of the Fox Hills deposits. The most important localities in the Ceratops beds are in Wyoming, especially in Converse County.

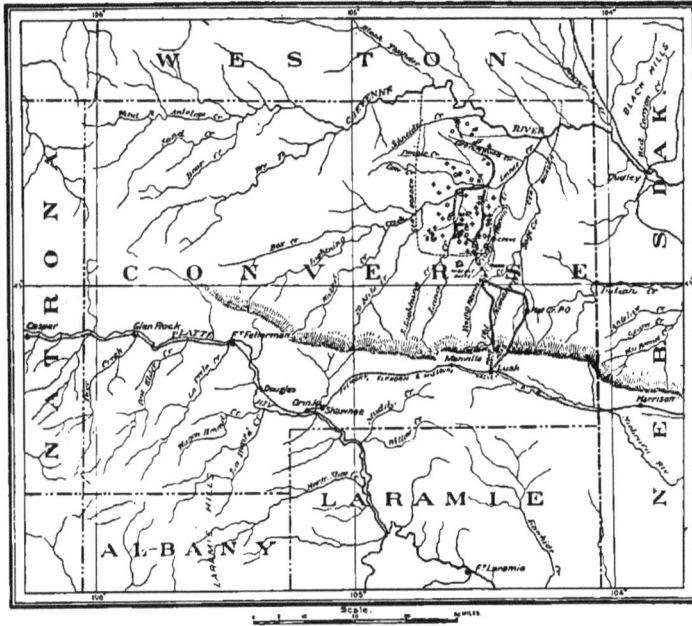

FIG. 53.—Map of Converse County, Wyoming; showing localities where skulls of the Ceratopsidæ have been discovered.

The position of each skull is indicated by a cross (+), and more than thirty of these specimens were found within the area bounded by the Cheyenne River and the dotted line. The localities given are based upon field notes made by Mr. J. B. Hatcher.

The fossils associated with the Ceratopsidæ are mainly dinosaurs, representing one or two orders and several families. Plesiosaurs, crocodiles, and turtles, of Cretaceous types, and many smaller reptiles, have left their remains in the same deposits. Numerous small mammals, also of ancient types, a few birds, and many fishes, are likewise entombed in this formation. Invertebrate fossils and plants are not uncommon in the same horizon.

TRICERATOPS.

THE SKULL.

The skull of Triceratops, the best-known genus of the family, has many remarkable features. First of all, its size, in the largest individuals, exceeds that of any land animal hitherto discovered, living or extinct, and is surpassed only by that of some of the cetaceans. The skull represented (one-ninth natural size) on Pl. LIX is one of the most perfect yet discovered. Those shown on Pl. LX, figs. 1–3, are both of comparatively young animals, but are about 6 feet in length. The type specimen of *Triceratops horridus* was an old individual, and the head, when complete, must have been 7 or 8 feet long. Two other skulls, nearly perfect, from the same horizon, have equal or still greater dimensions.

Another striking feature of the skull is its armature. This consisted of a sharp, cutting beak in front, a strong horn on the nose, a pair of very large pointed horns on the top of the head, and a row of sharp projections around the margin of the posterior crest. All these had a horny covering of great strength and power. For offense and defense they formed together an armor for the head as complete as any known. This armature dominated the skull, and in a great measure determined its form and structure. In some forms the armature extended over portions of the body.

The skull itself is wedge-shaped in form, especially when seen from above. The facial portion is very narrow, and much prolonged in front. In the frontal region the skull is massive and greatly strengthened, to support the large and lofty horn cores which formed the central feature of the armature. The huge, expanded, posterior crest, which overshadowed the back of the skull and neck, was evidently of secondary growth, a practical necessity for the attachment of the powerful ligaments and muscles that supported the head (Pl. LX, figs. 2 and 4).

THE ROSTRAL BONE.

The front part of the skull shows a very high degree of specialization, and the lower jaws have been modified in connection with it. In front of the premaxillaries there is a large, massive bone not before seen in any vertebrate, which the writer has named the rostral bone (*os rostrale*). It covers the anterior margin of the premaxillaries, and its sharp inferior edge is continuous with their lower border. This bone is much compressed, and its surface is very rugose, showing that it was covered with a strong, horny beak. It is a cartilage ossification, and corresponds to the predentary bone below.

The latter in Triceratops is also sharp and rugose, and likewise was protected by a strong, horny covering. The two together closely resemble the beak of some turtles, and as a whole must have formed a most powerful weapon of offense.

In one skull figured (Pl. LX, fig. 1) the rostral bone was free, and was not obtained. This was also true of the predentary bone and the nasal horn core. Hence these parts are represented in outline, taken from another specimen in which they are all present and in good preservation. In another skull represented (Pl. LIX, and Pl. LXI, figs. 1–3), the rostral bone and nasal horn core are in position and firmly coossified with the adjoining elements.

The premaxillary bones are large and much compressed transversely. Their inner surfaces are flat and meet each other closely on the median line. In old specimens they are firmly coossified with each other and with the rostral bone. Each sends upward a strong process to support the massive nasals. Another process, long and slender, extends upward and backward, forming a suture with the maxillary behind, and uniting in front with a descending branch of the nasal. The premaxillaries are much excavated externally for the narial aperture, and form its lower margin. They are entirely edentulous.

The maxillaries are thick, massive bones of moderate size, and subtriangular in outline when seen from the side. Their front margin is bounded mainly by the premaxillaries. They meet the prefrontal and lachrymal above, and also the jugal. The alveolar border is narrow and the teeth are small, with only a single row in use at the same time.

The nasal bones are large and massive, and greatly thickened anteriorly to support the nasal horn core. In two of the skulls figured these bones are separate, but in older individuals they are firmly coossified with each other and with the frontals. The nasal horn core ossifies from a separate center, but in adult animals it unites closely with the nasals, all traces of the connection being lost. It varies much in form in different species.

THE HORN CORES.

The frontal bones are quite short and early unite with each other and with the adjoining elements, especially those behind them. The frontal or central region of the skull is thus greatly strengthened to support the enormous horn cores which tower above. These elevations rest mainly on the postfrontal bones, but the supraorbitals and the postorbitals are also absorbed to form a solid foundation for the horn cores.

These horn cores are hollow at the base (Pl. LX, fig. 3), and in general form, position, and external texture agree with the corresponding parts of the Bovidæ. They vary much in shape and size in different species. They were evidently covered with massive, pointed horns, forming most powerful and effective weapons.

The orbit is at the base of the horn core, and is surrounded, especially above, by a very thick margin. It is oval in outline and of moderate size.

The postfrontal bones are very large, and meet each other on the median line. Posteriorly they join the squamosals and the parietals. At their union with the latter there is a median foramen (Pl. LX, fig. 3, x), which may correspond to the so-called "parietal foramen." In old individuals it is nearly or quite closed. When open it leads into a large sinus, extending above the brain case into the cavities of the horn cores. This foramen has not before been observed in dinosaurs.

THE POSTERIOR CREST.

The enormous posterior crest is formed mainly by the parietals, which meet the postfrontals immediately behind the horn cores. The posterior margin is protected by a series of special ossifications, which in life had a thick horny covering. These peculiar ossicles, which extend around the whole crest. the writer has called the epoccipital bones (Pl. LX, figs. 1–3, e. and Pl. LXI, fig. 8, e). In old animals they are firmly coossified with the bones on which they rest.

The lateral portions of the crest are formed by the squamosals, which meet the parietals in an open suture. Anteriorly they join the postfrontal elements, which form the base of the horn core, and laterally they unite with the jugals. The supratemporal fossæ lie between the squamosals and the parietals.

BASE OF SKULL.

The base of the skull has been modified in conformity with its upper surface. The basioccipital is especially massive, and strong at every point. The occipital condyle is very large, and its articular face nearly spherical, indicating great freedom of motion. The basioccipital processes are short and stout. The basipterygoid processes are longer and less robust.

The foramen magnum is very small, scarcely one-half the diameter of the occipital condyle. The brain cavity is especially diminutive, smaller in proportion to the skull than in any other known reptile.

The exoccipitals are also robust, and firmly coossified with the basioccipitals. They form about three-fourths of the occipital condyle, as in some of the chameleons. The supraoccipital is very small, and its external surface is excavated into deep cavities. It is coossified late with the parietals above and with the exoccipitals on the sides (Pl. LX, fig. 2).

The quadrate is robust and its head much compressed. The latter is held firmly in a deep groove of the squamosal. The anterior wing of the quadrate is large and thin, and closely united with the broad blade of the pterygoid.

The quadratojugal is a solid, compressed bone, uniting the quadrate with the large, descending process of the jugal. In the genus Tricera-tops the quadratojugal does not unite with the squamosal. In Cera-tops, which includes some of the smaller, less specialized forms of the family, the squamosal is firmly united to the quadratojugal by suture.

The quadratojugal arch in this group is strong and curves upward, the jugal uniting with the maxillary, not at its posterior extremity, but at its upper surface (Pl. LX, fig. 1). This greatly strengthens the center of the skull, which supports the horn cores, and also tends to modify materially the elements of the palate below. The pterygoids, in addition to their strong union with the quadrate, send outward a branch, which curves around the end of the maxillary.

The palatine bones are much smaller than the pterygoids. They are vertical, curved plates, outside and in front of the pterygoids, and uniting firmly with the maxillaries. The vomers join the pterygoids in front, where they appear as thin bones, closely applied to each other.

The transverse bones give some support to the maxillaries, which are further strengthened by close union with the pterygoids. They meet the pterygoids behind and the palatines in front.

THE LOWER JAWS.

The lower jaws show no specialization of great importance, with the exception of the predentary bone already described (Pl. LXI, figs. 4–6). There is, however, a very massive coronoid process rising from the posterior part of the dentary (Pl. LX, fig. 1). The articular, angular, and surangular bones are all short and strong, but the splenial is very long and slender, extending to the predentary. The angle of the lower jaw projects but little behind the quadrate.

The skull shown on Pl. LIX was discovered in the Ceratops beds of Wyoming by the writer's able assistant, Mr. J. B. Hatcher, who also found many other remains of dinosaurs.

THE BRAIN.

The brain of Triceratops appears to have been smaller in proportion to the entire skull than in any known vertebrate. Its relative size is shown on Pl. LXXVI, fig. 1.

The position of the brain in the skull does not correspond to the axis of the latter, the front being elevated at an angle of about 30° (Pl. LXI, fig. 7).

The brain case is well ossified in front, and in old animals there is a strong septum separating the olfactory lobes.

THE TEETH.

The teeth of Triceratops and its near allies are very remarkable in having two distinct roots. This is true of both the upper and lower series. These roots are placed transversely in the jaw, and there is a separate cavity, more or less distinct, for each of them. One of these teeth from the upper jaw, represented by several figures (Pl. LXI, figs. 9 and 10, and Pl. LXXVIII, fig. 4), is typical of the group.

The teeth form a single series only in each jaw. The upper and lower teeth are similar, but the grinding face is reversed, being on the inner side of the upper series and on the outer side of the lower series. The sculptured surface in each series is on the opposite side from that in use.

The teeth are not displaced vertically by their successors, but from the side. The crown of the young tooth, also with two strong roots, cuts its way between the alveolar margin and the adjacent root of the old tooth, but sometimes, as might be expected, advances between the two roots.

The teeth in this family are entirely confined to the maxillary and dentary bones. The rostral bone, the premaxillaries, and the predentary are entirely edentulous.

CERVICAL AND DORSAL VERTEBRÆ.

The atlas and axis of Triceratops are coossified with each other, and at least one other vertebra is firmly united with them. These form a solid mass, well adapted to support the enormous head (Pl. LXIV, fig. 1). The cup for the occipital condyle is nearly round and very deep. The rib of the second vertebra is coossified with it, but the third is usually free. The centrum of the fourth vertebra is free, and the remaining cervicals are of the same general form, all having their articular faces nearly flat.

The anterior dorsal vertebræ have very short centra, with flat articular ends, and resemble somewhat those of Stegosaurus, especially in the neural arch. This is shown in Pl. LXIV, figs. 3 and 4.

The posterior trunk vertebræ have also short, flat centra, but the diapophyses have faces for both the head and tubercle of the ribs, as in crocodiles, a feature but recently seen in dinosaurs.

THE SACRUM.

The sacrum was strengthened by union of several vertebræ, ten being coossified in one specimen of Triceratops (Pl. LXV). The middle or true sacral vertebræ have double transverse processes, diapophyses being present and aiding in supporting the ilium. This character has been seen hitherto in the Dinosauria only in Ceratosaurus and some other Theropoda.

The main support of the pelvis was borne by four vertebræ, which evidently constituted the original sacrum. In front of these, two others have only simple processes, and apparently were once dorsals or lumbars. Three vertebræ next behind the true sacrum have also single processes, and the fourth, or last of the series, has the rib process weak, and not reaching the ilium (Pl. LXV). Seen from the side the sacrum is much arched upward, and the neural spines of the true sacrum are firmly coossified. In the median region the sacral vertebræ have their centra much compressed, but the last of the series are widely expanded transversely. The whole appearance of the sacrum is remarkably avian. The neural canal of the sacral vertebræ has no special enlargement, thus differing widely from that in Stegosaurus.

THE CAUDAL VERTEBRÆ.

The caudal vertebræ are short and the tail was of moderate length. The first caudal has the anterior face of the centrum concave vertically, but flat transversely, and a short, massive neural spine with expanded summit (Pl. LXIV, figs. 5-7). In the median caudals the centra have biconcave articular faces and weak neural spines. The distal caudals are longer than wide, with the ends concave and nearly round.

THE SCAPULAR ARCH AND FORE LIMBS.

The scapula is massive, especially below. The shaft is long and narrow, with a thin edge in front and a thick posterior margin above the glenoid fossa. The distal portion has a median external ridge and a thick end (Pl. LXVI, fig. 1, *sc*).

The coracoid is rather small, and in old individuals may become united to the scapula. It is subrhombic in outline, and is perforated by a large and well-defined foramen. No indications of a sternum have yet been found in this group.

The humerus is large and robust and similar in form to that of Stegosaurus. In one individual it is nearly as long as the femur, proving that the animal walked on all four feet. The radius and ulna are comparatively short and stout, and the latter has a very large olecranon process, as shown in Pl. LXVI, fig. 3.

There were five well-developed digits in the manus. The metacarpals are short and stout, with rugose extremities. The distal phalanges are broad and hoof-like, showing that the fore feet were distinctly ungulate (Pl. LXIX).

THE PELVIS.

The pelvis in this group is very characteristic, and the three bones, ilium, ischium, and pubis, all take a prominent part in forming the acetabulum. The relative size and position of these are shown in Pl. LXVII, fig. 1, which represents the pelvic elements as nearly in the same plane as their form will allow, while retaining essentially their relative position in life.

The ilium is much elongated, and differs widely from that in any of the known groups of the Dinosauria. The portion in front of the acetabulum forms a broad, horizontal plate, which is continued backward over the acetabulum, and narrowed in the elongated, posterior extension. Seen from above, the ilium, as a whole, appears as a nearly horizontal, sigmoid plate. From the outside, as shown in the figure, the edge of this broad plate is seen.

The protuberance for the support of the pubis is comparatively small and elongated. The face for the ischium is much larger, and but little produced. The acetabular face of the ilium is quite narrow.

The pubis is massive, much compressed transversely, with its distal end widely expanded, as shown in the figures (Pl. LXVII). There is no true postpubis, but only a small postpubic process. The pubis itself projects forward, outward, and downward. Its union with the ilium is not a strong one, and is similar to that seen in the pubis of Stegosaurus.

The ischium is smaller than the pubis, but more elongate. Its shaft is much curved downward and inward, and in this respect it resembles somewhat the corresponding part of the pubis of the ostrich. There is no indication that the two ischia met closely at their distal ends, and they were probably united only by cartilage.

A comparison of this pelvis with that of Stegosaurus (Pl. LXXXI) shows some points of resemblance, but a wide difference in each of the elements. The pubis corresponds in its essential features to the prepubis of Stegosaurus, but the postpubis is represented only by a short process.

THE POSTERIOR LIMBS.

The femur is short, with the great trochanter well developed. The shaft is comparatively slender, and the distal end much expanded. The third trochanter is wanting, or represented only by a rugosity (Pl. LXVIII, fig. 1).

The tibia is of moderate length, and resembles that of Stegosaurus. The shaft is slender, but the ends are much expanded. The fibula is very slender, and the distal end was closely applied to the front of the tibia (Pl. LXVIII, fig. 2). In adult individuals the astragalus is firmly coossified with the distal end of the tibia, as in Stegosaurus.

The metatarsal bones which were functional are rather long, but massive. Their phalanges are stout, and the distal ones broad and rugose, indicating that the digits were terminated by very strong hoofs (Pl. LXIX, figs. 7–12).

All the limb bones and vertebræ in Triceratops and the nearly allied genera are solid.

THE DERMAL ARMOR.

Besides the armature of the skull, the body also in Triceratops was protected (Pl. LXX). The nature and position of the defensive parts in the different forms can not be determined with certainty, but various spines, bosses, and plates have been found that clearly pertain to the dermal covering of Triceratops, or nearly allied genera. Several of these ossifications were probably placed on the back, behind the crest of the skull, and some of the smaller ones may have defended the throat, as in Stegosaurus.

TOROSAURUS.

In the type specimen on which this genus was based the greater portion of the skull is preserved, and this presents so many points of interest that a figure of it, one-twentieth natural size, is here given in Pl. LXII, fig. 1. The second species is represented also by the skull, which, although not complete, supplements the type in several important respects, and figures of its posterior portions are likewise given in the same plate and in fig. 54, on the opposite page. Both specimens are of gigantic size, one skull measuring 5½ feet across the parietal crest, and the other is nearly as large. They differ widely, moreover, from the huge horned dinosaurs hitherto found in the same general horizon, and in the skull present characters of much interest.

THE SKULL.

In *Torosaurus latus*, the species first described, the skull appears wedge-shaped when seen from above, as shown in Pl. LXII. The facial portion is very short and pointed, and somewhat snilline in form. The nasal horn core is compressed, with a sharp apex directed forward. The frontal horn cores are large and strongly inclined to the front, extending apparently in advance of the nasal protuberance. The long, slender squamosals diverge rapidly as they extend backward, their outer margins being nearly on a line with the facial borders in the maxillary region.

The parietal forms more than half of the upper surface of the skull, and is the most characteristic element in its structure. In the posterior part are two very large apertures, oval in outline, with their outer margin at one point formed by the squamosal. The rest of the border is thin and somewhat irregular, showing that the openings are true

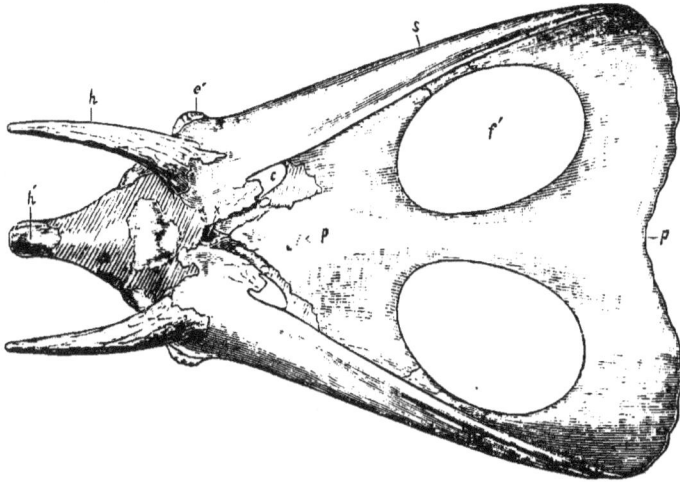

FIG. 54.—Skull of *Torosaurus gladius* Marsh: seen from above. One-twentieth natural size.
c, supratemporal fossa; *e'*, epijugal bone; *f'*, parietal fontanelle; *h*, horn core; *h'*, nasal horn core; *p*, parietal; *s*, squamosal; *x*, pineal foramen (?).

fontanelles. This is still better seen in the second species represented in the same plate, fig. 2, and in fig. 54, above. In the latter specimen, however, these vacuities are entirely in the parietal, a thin strip of bone separating them on either side from the squamosal. A second pair of openings, much smaller, apparently the true supratemporal fossæ, are shown in the type specimen. These are situated mainly between the parietal and squamosal, directly behind the bases of the large horn cores (Pl. LXII, fig. 1, *c*). The same apertures are represented in the genus Triceratops by oblique openings, as in the skull shown on Pl. LX, fig. 3, *c*, where the front border of each is formed by the postfrontal.

Between these openings, in the type of Torosaurus, is a third pair of apertures (Pl. LXII, fig. 1, *e'*). These are quite small, nearly circular in outline, and entirely in the parietal, although probably connected originally with the supratemporal fossæ. Another pair of still smaller foramina may be seen in the same skull, close to the median line, and separated from each other by the anterior projection of the parietal. A deep groove leads forward to each of these foramina, along the suture between the parietal and the postfrontal. The position and direction of these perforations suggest that they may correspond to the foramen seen in Triceratops, and shown on Pl. LX, fig. 3, *x*. The same foramen is shown above in fig. 54.

The extreme lightness and great expanse of the posterior crest in Torosaurus make it probable that it was encased in the integuments of the head, and that no part of it was free. The outer borders of both the parietal and the squamosals show no marginal ossifications, as in the other known genera of the group, but the presence of a large, separate, epijugal bone in one specimen suggests that epoccipitals may yet be found.

The open perforations in the parietal, which have suggested the name Torosaurus, readily separate this genus from all the gigantic species hitherto known in the Ceratopsidæ, but may perhaps be found in some of the smaller and less specialized forms, from lower horizons of the same formation.

CERATOPS.

The genus Ceratops so far as at present known is represented by individuals of smaller size than those of Triceratops, and in some instances, at least, of quite different proportions. The type specimen is shown in Pl. LXIII. A third genus, Sterrholophus, can be readily distinguished from the other two by the parietal crest, which had its entire posterior surface covered with the ligaments and muscles supporting the head. In Ceratops and Triceratops a wide margin of this surface was free, and protected by a thick, horny covering. There is some evidence that still other forms, quite distinct, left their remains in essentially the same horizon, but their true relation to the above genera can not be settled without further discoveries.

STERRHOLOPHUS.

With the successive changes in the parietal in the Ceratopsidæ, there were corresponding variations in the squamosals, and these bones also will serve to distinguish the principal genera from one another. In Pl. LXIII the squamosals of three genera of this group are shown, and the wide difference between them, when seen from the inside, is especially noticeable. In fig. 4 of this plate, the long, slender, right squamosal of Torosaurus, with its smooth outer border, is well represented. In fig. 5 is seen the same bone of Sterrholophus, with a serrate outer margin and smooth inner surface, also shown in fig. 1 of Pl. LX. Next, in fig. 6 is the small, short squamosal of Ceratops, nearly bisected by its deep quadrate groove. The free sculptured border of both the parietal and squamosals of Triceratops is clearly shown in Pl. LX, fig. 4, where the contrast with the corresponding parts in fig. 2 is noteworthy.

AGATHAUMAS.

Three other generic names have been applied by Cope to remains of
Ceratopsidæ found in this country, namely, Agathaumas, Polyonax,
and Monoclonius. The first of these was based on part of a skeleton,
without the skull, found in Wyoming. The second name was given
to various fragments from Colorado, including parts of horn cores,
regarded as ischia, but all these may be the same generically as the
preceding specimen. The third name, Monoclonius, was used for a
skeleton from Montana, with parts of the skull and teeth preserved.
This animal was one of the smallest of the group, while the other
remains pertained to reptiles of larger size, but not of the gigantic
proportions of those more recently described. So far as can be judged
from the descriptions and figures of the type specimens, the three
generic names just cited can not be used for any of those previously
mentioned in this article. A comparison of the principal characters
will place this beyond reasonable doubt.

In the type of Agathaumas the remains best preserved are from the
pelvic region, which, according to Cope,[1] possesses the following features: The ilium has no facet nor suture for the pubis at the front of
the acetabulum, and the base of the ischium is coossified with the ilium.
There are eight, or perhaps nine, sacral vertebræ, with the neural
spines of the first five mere tuberosities. The diapophyses are in pairs,
and the last sacral vertebra is reduced and elongate. These characters, and some others found in the description cited, are certainly distinctive, but do not apply to any of the allied fossils described by the
writer. Portions of the type specimen, moreover, are in the Yale
museum, as well as other remains from near the same locality. The
fossils described as Polyonax, and other similar specimens collected in
the same region, afford at present no evidence for separation from
Agathaumas.

MONOCLONIUS.

The small dinosaur for which the name Monoclonius was proposed
is perhaps generically distinct from Agathaumas, but no conclusive
evidence of this has yet been presented. The description given makes
the teeth, dorsal vertebræ, and pelvis different from those of any of
the larger forms, and the T-shaped parietal (figured first by Cope as
an episternal bone) is especially distinctive. None of the other known
Ceratopsidæ have the parietal fontanelles except Torosaurus, one of
the most gigantic forms discovered, and this genus differs from Monoclonius, as described, in various important points. The very long
frontal horn cores, directed forward, the narrow, elongate squamosals,
the absence of a median crest on the parietal, as well as the form and
anterior connections of this bone, all serve to distinguish clearly the
former from the latter.

[1] Vertebrata of the Cretaceous formations of the West, p. 53, 1875.

RESTORATION OF TRICERATOPS.

PLATE LXXI.

The abundant material now available for examination makes it possible to attempt a restoration of one characteristic form of this group, and the result is given in Pl. LXXI. This figure, about one-fortieth natural size, is reduced from a large outline plate of a memoir now in preparation by the writer for the United States Geological Survey.

This restoration is based mainly on two specimens. One of these is the type of *Triceratops prorsus* Marsh, in which the skull, lower jaw, and cervical vertebræ are in remarkable preservation. The other specimen, although somewhat larger, is referred to the same species. It consists of parts of the skull, of vertebræ, the pelvic arch, and nearly all the important limb bones. The remaining portions are taken mostly from other remains found in the same horizon and localities, and at present are not distinguishable specifically from the two specimens above mentioned. The skull as here represented corresponds in scale to the skeleton of the larger individual.

In this restoration the animal is represented as walking, and the enormous head is in a position adapted to that motion. The massive fore limbs, proportionally the largest in any known dinosaur, correspond with the head, and indicate slow locomotion on all four feet.

The skull is, of course, without its strong horny covering on the beak, horn cores, and posterior crest, and hence appears much smaller than in life. The neck seems short, but the first six cervical vertebræ are entirely concealed by the crest of the skull, which in its complete armature would extend over one or two vertebræ more. The posterior dorsals with their double-headed ribs continue back to the sacrum itself, there being no true lumbars, although two vertebræ, apparently once lumbars, are now sacrals, as their transverse processes meet the ilia, and their centra are coossified with the true sacrum. The four original sacral vertebræ have their neural spines fused into a single plate, while the posterior sacrals, once caudals, have separate spines directed backward.

No attempt is made in this restoration to represent the dermal armor of the body, although in life the latter was more or less protected. Various spines, bosses, and plates, indicating such dermal armature, have been found with remains of this group, but the exact position of these specimens can be, at present, only a matter of conjecture.

This restoration gives a correct idea of the general proportions of the entire skeleton in the genus Triceratops. The size, in life, would be about 25 feet in length and 10 feet in height.

DISTINCTIVE CHARACTERS OF GROUP.

This group so far as at present investigated is very distinct from all other known dinosaurs, and whether it should be regarded as a family,

Ceratopsidæ, as first described by the writer, or as a suborder, Ceratopsia, as later defined by him, will depend upon the interpretation and value of the peculiar characters manifested in its typical forms.

The main characters which separate the group from other families of the Dinosauria are as follows:

(1) A rostral bone, forming a sharp, cutting beak.

(2) The skull surmounted by massive horn cores.

(3) The expanded parietal crest, with its marginal armature.

(4) A pineal foramen (?).

(5) The teeth with two distinct roots.

(6) The anterior cervical vertebræ coossified with each other.

(7) The dorsal vertebræ supporting, on the diapophysis, both the head and tubercle of the rib.

(8) The lumbar vertebræ wanting.

The animals of this group were all herbivorous, and their food was probably the soft, succulent vegetation that flourished during the Cretaceous period. The remains here figured are from the Ceratops beds of the upper Laramie, on the eastern slope of the Rocky Mountains.

The only known European member of this group is the Struthiosaurus Bunzel, 1871, apparently identical with Crataeomus Seeley, 1881. It is from the Gosau formation of Austria, and the locality was visited by the writer in 1864. Although only fragments, mostly of the skeleton and dermal armor, are known, some of these are very characteristic. One specimen figured by Seeley, and regarded as a dermal plate bearing a horn-like spine, is certainly part of the skull. It is very similar in form to some of the horn cores of the smaller species of Ceratops.

CLAOSAURIDÆ.

The next most important family of herbivorous dinosaurs from the Cretaceous of North America is the Claosauridæ, and of these the type genus is Claosaurus, described by the writer in 1890, from a specimen found by him in Kansas in 1872. Several fortunate discoveries since made have rendered this genus one of the best known of American forms, and hence the principal characters of the skull and skeleton are here given in detail.

CLAOSAURUS.

THE SKULL.

The skull of Claosaurus is long and narrow, with the facial portion especially produced. The anterior part is only moderately expanded transversely. Seen from the side (Pl. LXXII, fig. 1), the skull shows a blunt, rugose muzzle, formed above by the premaxillary and below by the predentary, both probably covered in life with a thick, corneous integument.

Behind the upper part of this muzzle is an enormous lateral cavity, which includes the narial orifice, but was evidently occupied in life mainly by a nasal gland, somewhat like that in the existing monitor, and seen also in some birds. This cavity is bounded externally by the nasal bone and the premaxillary. The median septum between the two narial orifices was only in part ossified, the large oval opening now present in the skull probably having been closed in life by cartilage.

The orbit is very large and subtriangular in outline. It is formed above by the prefrontal, frontal, and postfrontal, and below mainly by the jugal. There are no supraorbital bones. A distinct lachrymal forms a portion of the anterior border. The infratemporal fossa is large, and is bounded above by the postfrontal and squamosal, and below by the jugal. The quadrate forms a small portion of the posterior border.

Seen from in front (Pl. LXXII, fig. 2), the skull of Claosaurus is subovate in outline, with the narrow portion above. The premaxillaries and the predentary bone forming the rugose muzzle are especially massive and prominent, and the powerful lower jaws seem out of proportion to the more delicate bones of the cranium.

Seen from above (Pl. LXXII, fig. 3), the structure of the skull itself is shown to the best advantage. In front are the large premaxillaries, deeply excavated for the nasal openings. These bones are separate, and each sends back a long, slender process inside the anterior projection of the nasal, and a still longer process forming the lower border of the narial orifice, and extending to the lachrymal. The front of the premaxillaries is especially massive, and its surface rugose, indicating that it had been covered with a horny beak. The lower border is sharp, conforming to the corresponding surface of the predentary bone, which was doubtless also inclosed in a horny covering. The premaxillaries were entirely without teeth.

The nasal bones are long and slender, and especially produced in front, where they embrace the posterior median extensions of the premaxillaries. They also meet the lateral processes of the premaxillaries behind the nasal openings, and likewise touch the lachrymals. Farther back they meet the prefrontals and closely unite with the frontals, as shown in Pl. LXXII.

The frontal bones are quite short, and nearly as wide as long. They are united to each other by a well-marked suture. Their upper surface is smooth, and there is a slight depression on either side, posterior to the suture with the prefrontals. Each frontal bone forms a portion of the upper border of the orbit, and behind this meets the postfrontal. Posteriorly the frontals form the anterior border of the supratemporal fossæ, and between these unite by suture with the coossified parietals.

The latter bones are quite small, and appear on the upper surface of the skull mainly as a narrow ridge separating the supratemporal fossæ, and ending behind in a point, between the median processes of the squamosals. The parietals expand below, where they cover the posterior portion of the brain cavity.

The squamosal bones are robust, and their position and connections are well shown on Pl. LXXII, figs. 1 and 3. On the median line above they meet the narrow extension of the parietals, and exterior to this they form the posterior borders of the supratemporal fossæ. In front they unite by a strong process with the posterior branch of the postfrontals. Their posterior border is joined mainly to the exoccipitals. On the outer surface of each squamosal there is a deep pit to receive the head of the quadrate, and in front of this a short, narrow process extends down the quadrate, forming a part of the border of the infratemporal fossa.

The quadrate bone and its main connections are shown on Pl. LXXII, figs. 1–3. It is firmly supported above by the squamosal, but its distinct, rounded head indicates the possibility of some motion. On the outer surface in front it joins by open suture the strong jugal bone, and below this unites with the small, discoid quadratojugal. Its inner margin extends forward into a broad, thin wing for union with the pterygoid. The lower extremity is massive, and moderately expanded transversely for articulation with the lower jaw.

The jugal is one of the most characteristic parts of the skull, as may be seen from the figures on Pl. LXXII. Its main portion is robust, much compressed, and convex externally. On its upper margin it forms the lower border of the orbit and of the infratemporal fossa, sending up a strong process between them, which extends inside and in front of the postorbital branch of the postfrontal. In front it is strongly united to the maxillary, and above joins by suture with the lachrymal.

The maxillary bone in Claosaurus is of moderate dimensions, and seen from the outside is overshadowed by the premaxillary and jugal, as shown in the same plate, fig. 1. Its lower dentary border is thickly studded with a regular series of teeth, which slightly overlap those of the lower jaw. From above only a small portion of the maxillary is visible, as seen in Pl. LXXII, fig. 3.

The lower jaws are long and massive. The predentary bone is robust, and especially fitted for meeting the strong beak above. The dentary bones are large and powerful, with elevated coronoid processes. The angular and surangular bones are, however, quite short and not especially strong.

THE TEETH.

The teeth of Claosaurus are confined entirely to the maxillary and dentary bones. In each the teeth are very numerous, and are arranged in vertical series, so that they succeed each other as the functional teeth are worn away. This is seen in Pl. LXXVIII, fig. 2, which shows the form of the teeth and their relations to each other in the same series. The number of teeth in each depends upon the position, the series near the middle of the jaw having the greatest number, sometimes six or more. The teeth of the upper jaw have the external face of the crown covered with enamel and ridged. In the lower jaw this

is reversed, the ridged face of the crown being on the inside. This
arrangement greatly increased the cutting power of the jaws. The
food was probably soft vegetation.

THE BRAIN.

The brain of Claosaurus was very small, its size in proportion to the
skull being represented in Pl. LXXVI, fig. 2, which also shows the
exact position of the brain in the cranium. A cast of the brain cavity
is shown in Pl. LXXVII, fig. 3, one-fourth natural size. The brain as
a whole was considerably elongated, especially the posterior half. The
olfactory lobes were well developed, and not separated by an osseous
septum. The cerebral hemispheres were comparatively large, forming
nearly or quite half the entire brain. The optic lobes were narrow,
but considerably elevated. The cerebellum was rather small, and also
much compressed. The medulla was of good size, and nearly circular
in transverse outline. The pituitary body was quite large. The interpretation of some of the more minute features of the brain is a matter
of difficulty, and will be more fully discussed elsewhere.

THE VERTEBRÆ.

The main characters of the vertebral column of Claosaurus are well
shown in the restoration (Pl. LXXIV). There are thirty vertebræ
between the skull and sacrum, nine in the sacrum, and about sixty in
the tail. The whole vertebral column was found in position except the
terminal caudals, which are here represented in outline. The cervical
vertebræ are strongly opisthocœlian, and the first eleven have short
ribs. The dorsals are also opisthocœlian. There are no true lumbar
vertebræ, as the last of those in front of the sacrum support free ribs.
The anterior caudals are opisthocœlian. The first and second have no
chevrons. Behind these the chevron bones are very long, indicating a
powerful, compressed tail, well adapted for swimming.

In the median dorsal region, between the ribs and the neural spines,
are numerous rod-like ossified tendons, which increase in number in the
sacral region and along the base of the tail, and then gradually dimin-
ish in number and size, ending at about the thirty-fifth caudal. These
ossified tendons are well shown in the restoration, and are of much
interest. They are not unlike those in Iguanodon described by Dollo,
but as a rule are more elongate, and appear to lack the definite arrangement in rhomboidal figures observed in that genus.

THE FORE AND HIND LIMBS.

The fore limbs are unusually small in comparison with the posterior,
and the relative size of the two is shown on Pl. LXXIII. The scapular
arch presents many points of interest. The scapula is large, and so
much curved that its shaft is nearly at right angles to the articular
faces of its lower extremity (Pl. LXXIII, fig. 1, s). On the anterior

margin, above the articulation for the coracoid, is a strong protuberance, with a well-defined facet, adapted to the support of the clavicle,
if such a bone were present. The coracoid is very small, and is perforated by a large foramen. The two peculiar bones now generally
regarded as belonging to the sternum were separate, as shown in Pl.
LXXV, fig. 4.

The humerus is comparatively short, and has a prominent radial
crest. The radius and ulna are much elongated. the latter being longer
than the humerus. and the radius about the same length. The ulna
has a prominent olecranon process, and is a stouter bone than the
radius. The carpal bones were quite short. and appear to have been
only imperfectly ossified. The fore foot, or manus, was very long, and
contained three functional digits only. The first digit was rudimentary,
the second and third were nearly equal in length, the fourth was shorter
and less developed, and the fifth entirely wanting. as shown in Pl.
LXXIII, fig. 1.

In the functional digits (II, III, IV) the phalanges are elongate, thus
materially lengthening the fore foot. The terminal phalanges of these
digits are broad and flat, showing that they were covered with hoofs,
and not with claws. The limb as a whole was thus adapted to loco-

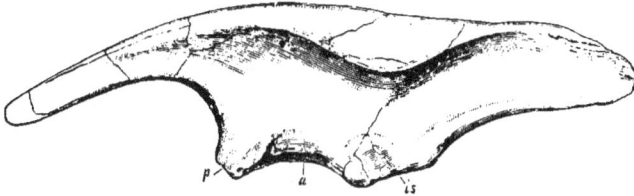

Fig. 55.—Ilium of *Claosaurus agilis* Marsh; seen from the left. One-sixth natural size
a, acetabular border; *is*, face for ischium; *p*, face for pubis.

motion or support, and not at all for prehension, although this might
have been expected from its small size and position.

The elongation of the forearm and manus is a peculiar feature, especially when taken in connection with the ungulate phalanges. It may,
perhaps, be explained by supposing that the animal gradually assumed
a more erect position until it became essentially a biped, while the fore
limbs retained in a measure their primitive function, and did not become
prehensile as in some allied forms.

The pelvis is shown in Pl. LXXIII. figs. 2 and 3. and has been
fully described by the writer. Its most notable features are seen in
the pubis and ischium, the former having a very large expanded prepubis, with the postpubis rudimentary, while the shaft of the ischium
is greatly elongated. The ilium of the type species is shown in fig. 55.

The femur is long, and the shaft nearly straight. The great trochanter
is well developed. while the third trochanter is large and near the middle
of the shaft, as shown in Pl. LXXIII, fig. 2. The external condyle of

the distal end is projected well backward, indicating great freedom of motion at the knee.

The tibia is shorter than the femur and has a prominent cnemial crest. The distal end is much flattened, and the astragalus is closely adapted to it. The fibula is very straight, with its lower end flattened and closely applied to the front of the tibia. The calcaneum is large, with its concave upper surface closely fitted to the end of the fibula. Of the second row of tarsals only a single one appears to be ossified, and that is very small and thin, and placed between the calcaneum and the fourth metatarsal, nearly or quite out of sight.

The hind foot, or pes, had but three digits, the second, third, and fourth, all well developed and massive. The terminal phalanges were covered with broad hoofs. The first and fifth digits were entirely wanting.

A comparison of the limbs and feet of Claosaurus, as here described and figured, with those of three allied forms from the Jurassic, Stegosaurus, Laosaurus, and Camptosaurus, as shown on Pls. XLVIII, LIV, and LV, is especially instructive. These three genera have already been quite fully described and figured by the writer, but new points of interest have been made out by the recent investigation of more perfect material. The present figures will show more accurately some of the mutual relations of these early herbivorous dinosaurs to one another, as well as to their successors in Cretaceous time. The gradual changes that can be traced from one to the other will be discussed in a later communication.

All the limb bones in Claosaurus are solid, thus distinguishing it from Trachodon (Hadrosaurus). The separate ischium not coossified with the pubis, the absence of a fourth digit in the hind foot, and other marked characters, also make the genus distinct from Pteropelyx, the skull of which is not known.

RESTORATION OF CLAOSAURUS.

PLATE LXXIV.

The reptile here restored was nearly 30 feet in length when alive, and about 15 feet in height in the position represented in Pl. LXXIV. The remains were obtained in the Ceratops beds of the Laramie, in Wyoming. Among the associated fossils are the gigantic Triceratops and Torosaurus, which were also herbivorous dinosaurs, and with them were found the diminutive Cretaceous mammals recently described by the writer.

TRACHODONTIDÆ.

The genus Trachodon of Leidy, which has been admirably described under the name Hadrosaurus by that author,[1] is a near ally of Claosaurus, but quite distinct. The generic name Diclonius Cope should be regarded as a synonym of Trachodon. The teeth of one species of this genus are shown in Pl. LXXV, figs. 1 and 2.

[1] Cretaceous Reptiles of the United States. p. 76, 1865.

PALÆOSCINCUS.

A new reptilian genus and species, *Palæoscincus costatus*, was proposed by Dr. Leidy, in 1856, for a single tooth found by Dr. Hayden in the Judith Basin. This tooth was more fully described and figured by Leidy in 1859. The specimen showed well-marked characters, and many similar teeth have since been found, both in the Judith Basin and in various other localities of the same horizon.

A smaller species, apparently of the same genus, is not uncommon in the Ceratops beds of Wyoming, and a characteristic tooth is shown on Pl. LXXV, fig. 3. This is the type specimen of the species *Palæoscincus latus*. The crown of the tooth in this species is broader and the apex more pointed than in the first species described, and this is clearly shown in comparing the present figures on Pl. LXXV with those given by Leidy.

NODOSAURIDÆ.

NODOSAURUS.

Another genus of Stegosauria, from a lower horizon in the Cretaceous, was discovered several years since, in Wyoming, and the type specimen is now in the Yale museum. This genus, Nodosaurus, was described by the writer in 1889. The skull is not known, but various portions of the skeleton were secured. One characteristic feature in this genus is the dermal armor, which appears to have been more complete than in any of the American forms hitherto found. This armor covered the sides closely, and was supported by the ribs, which were especially strengthened to maintain it. In the present specimen portions of it were found in position. It was regularly arranged in a series of rounded knobs in rows, and these protuberances have suggested the generic name.

Near the head the dermal ossifications were quite small, and those preserved are quadrangular in form, and arranged in rows. The external surface is peculiarly marked by a texture that appears interwoven, like a coarse cloth. This has suggested the specific name, and is well shown in Pl. LXXV, fig. 5.

The fore limbs are especially massive and powerful, and are much like those of the Jurassic Stegosaurus. There were five well-developed digits in the manus, and their terminal phalanges are more narrow than usual in this group. The ribs are T-shaped in transverse section, and thus especially adapted to support the armor over them. The caudal vertebræ are more elongate than those of Stegosaurus, and the middle caudals have a median groove on the lower surface of the centrum.

The animal when alive was about 30 feet in length. The known remains are from the middle Cretaceous of Wyoming.

16 GEOL, PT 1——15

DISTRIBUTION OF ORNITHOPODA.

The great group which the author has called the Ornithopoda is well represented in Europe by Iguanodon and its allies. The remarkable discoveries in the Wealden of Belgium of a score or more skeletons of Iguanodon have furnished material for an accurate study of the genus which they represent, and, indirectly, of the family. The genus Iguanodon, founded by Mantell in 1824, is now the best known of European forms, while Hypsilophodon Huxley, 1870, also from the Wealden, is well represented, and its most important characters are fully determined. For comparison with American forms, restorations of both Hypsilophodon and Iguanodon are given on Pls. LXXXIII and LXXXIV. The other genera of this group, among which are Mochlodon Bunzel, 1871, Vectisaurus Hulke, 1879, and Sphenospondylus Seeley, 1883, are described from less perfect material, and further discoveries must decide their distinctive characters.

None of these genera are known from America, but allied forms are not wanting. A distinct family, the Trachodontidæ, is especially abundant in the Cretaceous, and another, the Camptosauridæ, includes most of the Jurassic species. The latter are the American representatives of the Iguanodontidæ. The nearest allied genera are, apparently, Iguanodon and Camptosaurus for the larger forms, and Hypsilophodon and Laosaurus for those of small size. A few isolated teeth from each country suggest that forms more nearly related may at any time be brought to light.

Many generic names have been proposed for members of this group found in America and in Europe, but in most cases they are based on fragmentary, detached specimens, which must await future discoveries before they can be assigned to their true place in the order.

In conclusion, it may be said that the three great groups of Dinosauria are each well represented in Europe as well as in America. Some of the families, also, of each order have representatives in the two regions, and future discoveries will doubtless prove that others occur in both.

No genera common to the two continents are known with certainty, although a few are so closely allied that they can not be distinguished from one another by the fragmentary specimens that now represent them.

From Asia and Africa, also, a few remains of dinosaurs have been described, and the latter continent promises to yield many interesting forms. Characteristic specimens, representing two genera, one apparently belonging to the Stegosauria, and one to the Theropoda, are already known from South Africa, from the region so rich in other extinct Reptilia.

From Australia no Dinosauria, except a single specimen, have as yet been recorded, but many more will undoubtedly be found there, as reptiles of this great group were the dominant land animals of the earth during all Mesozoic time.

PART IV.

CONCLUSION.

The brief review of North American Dinosaurs given in the preceding pages, in connection with the accompanying illustrations, will make the reader acquainted with the more important type specimens of this interesting group of reptiles, as now known from this continent. To discover and bring together these remains, representing several hundred individuals, from widely separated localities and various geological horizons, has been a long and laborious undertaking, attended with much hardship, and often with danger, but not without the pleasure that exploration in new fields brings to its votaries. These researches, especially in the West, have been continued by the writer more than a score of years, and have led him across the Rocky Mountains a still greater number of times. The field work thus prosecuted has been of great service in the subsequent study of the specimens secured, especially in determining the natural position in life of each animal investigated.

In comparing the type specimens of these various animals, one with another, as they were found and as they appeared when removed from the vesture of their entombment, many questions have suggested themselves that can not be answered in the present limited paper. Resemblances and differences are striking, both in structure and form, in these ancient reptiles, but the true meaning of such features is a difficult problem to solve. On the interpretation of characters thus exhibited in these animals depend both the laws of their classification and theories of their origin.

COMPARISON OF CHARACTERS.

In the concluding part of the present paper a number of plates (LXXVI-LXXXI) have been given with a view to illustrate especially the corresponding parts of various animals of different orders, showing the wide divergence in some points of nearly allied forms, and the approach in particular features of types clearly distinct.

In Pl. LXXVI four skulls of as many typical genera of herbivorous dinosaurs (Triceratops, Claosaurus, Camptosaurus, and Diplodocus) are represented, with a cast of the brain cavity of each in position. All are so drawn that they can be readily compared, thus exhibiting in a striking manner both the diminutive size of the brain in each in proportion to that of the skull, and also the form of the brain cavity, when seen from above. In the next plate (LXXVII) the brain casts alone of several dinosaurs, as seen from the side, are exhibited, and with them for comparison the corresponding cast of a young alligator. The special features of the dinosaur brain are well shown in these two plates.

Pl. LXXVIII will make clear the wide divergence of forms of teeth in four different families of predentate dinosaurs. The typical genera,

Camptosaurus, Claosaurus, Stegosaurus, and Triceratops, each have teeth of a distinct type, yet it seems possible to trace the gradation of one to the other through different intermediate forms. In Triceratops the teeth have two distinct roots, a feature unknown in any other reptiles, living or extinct, but characteristic of mammals.

The series of pubic bones of herbivorous dinosaurs shown on Pl. LXXIX is especially instructive, as they indicate how the anterior and posterior elements of the pubis may vary in the Predentata, and thus afford good characters for classification. The same is true, but in a less degree, of the ischia represented on Pl. LXXX, which all pertain to one group of dinosaurs. The comparison may even be carried much further, as in the two other orders (the Theropoda and Sauropoda) some families have ischia of the type here represented, as shown on Pls. XXVIII and XXXV.

The pelves represented on Pl. LXXXI, pertaining to the three predentate genera, Camptosaurus, Triceratops, and Stegosaurus, will supplement the facts presented on the two preceding plates. The series might be much further extended, and prove equally instructive. This will be done by the writer in the monographs now in preparation, as in these the whole subject of dinosaurian reptiles will receive careful consideration.

RESTORATIONS OF EUROPEAN DINOSAURS.

The remaining restorations of dinosaurs in this paper are four in number, and represent some of the best-known European forms, types of the genera Compsognathus, Scelidosaurus, Hypsilophodon, and Iguanodon. These outline restorations have been prepared by the writer mainly for comparison with the corresponding American forms, but in part to insure, so far as the present opportunity will allow, a more comprehensive review of the whole group. The specimens restored are all of great interest in themselves, and of special importance when compared with their nearest American allies.

COMPSOGNATHUS.

PLATE LXXXII.

The first restoration, that of *Compsognathus longipes* Wagner, 1861, shown one-fourth natural size on Pl. LXXXII, is believed to represent fairly well the general form and natural position, when alive, of this diminutive carnivorous dinosaur that lived during the Jurassic period. The basis for this restoration is (1) a careful study of the type specimen itself, made by the writer in Munich in 1881; (2) an accurate cast of this specimen, sent to him by Professor von Zittel; and (3) a careful drawing of the original, made by Krapf in 1887. The original description and figure of Wagner (Bavarian Academy of Sciences, 1861), and those of later authors, have also been used for some of the details.

No restoration of the skeleton of this unique dinosaur has hitherto been attempted.[1]

Compsognathus has been studied by so many anatomists of repute since its discovery that any attempt to restore the skeleton to a natural position will be scrutinized from various points of view. Interest in this unique specimen led the writer long ago to examine it with care, and he has since made a minute study of it, as related elsewhere, not merely to ascertain its anatomy, but also to learn, if possible, what its relations are to another diminutive form, Hallopus, from a lower horizon in America, which has been asserted to be a near ally. Both are carnivorous dinosaurs, probably, but certainly on quite different lines of descent.

The only previous attempt to restore this remarkable dinosaur was by Huxley when in America in 1876. He made a rapid sketch from the Wagner figure, and this was enlarged for his New York lecture. This sketch represents the animal sitting down, a position which such dinosaurs occasionally assumed, as shown by the footprints in the Connecticut Valley, which Huxley examined in place at several localities with great interest.

In the present restoration of Compsognathus (Pl. LXXXII) the writer has tried to represent the animal as walking in a characteristic lifelike position.

SCELIDOSAURUS.

PLATE LXXXIII.

The second of these restorations is that of *Scelidosaurus Harrisonii* of Owen, shown one-eighteenth natural size on Pl. LXXXIII. This reptile was an herbivorous dinosaur of moderate size, related to Stegosaurus, and was its predecessor from a lower geological horizon in England. This restoration is essentially based upon the original description and figures of Owen (Palæontographical Society, 1861). These have been supplemented by the writer's own notes and sketches, made during examinations of the type specimen now in the British Museum.

Scelidosaurus is a near relative, as it were, of one of the American forms, Stegosaurus, now represented by so many specimens that the skull, skeleton, and dermal armor are known with much certainty. The English form usually called Omosaurus is still more nearly allied to Stegosaurus, perhaps identical with it.[2]

A restoration of the skeleton of Scelidosaurus by Dr. Henry Woodward will be found in the British Museum Guide to Geology and Palæontology, 1890, p. 19. The missing parts are restored from Iguanodon, and the animal is represented as bipedal, as in that genus.

[1] The remains of the embryo within the skeleton of Compsognathus, first detected by the writer in 1881 while examining the type specimen, is not represented in the present restoration. This unique fossil affords the only conclusive evidence that dinosaurs were viviparous.

[2] The generic name Omosaurus was preoccupied by Leidy in 1856.

In the present outline restoration of Scelidosaurus the writer has endeavored merely to place on record his idea of the form and position of the skeleton when the animal was alive, based on the remains he has himself examined. In case of doubt, as, for example, in regard to the front of the skull, which is wanting in the type specimen, a dotted outline is used, based on the nearest allied form. Of the dermal armor, only the row of plates best known is indicated. The position chosen in this figure (Pl. LXXXIII) is one that would be assumed by the animal in walking on all four feet, and this is believed to have been its natural mode of progression.

HYPSILOPHODON.

PLATE LXXXIV.

The third of these restorations, that of *Hypsilophodon Foxii* Huxley, 1870, given in outline one-eighth natural size on Pl. LXXXIV, has been made with much care, partly from the type specimen, and in part from other material mostly now in the British Museum. The figures and description by the late Dr. Hulke[1] were of special value, although the conclusions of the writer as to the natural position of the animal when alive do not coincide with those of his honored friend, who did so much to make this genus of dinosaurs, and others, known to science. The restoration by Dr. Hulke represented the animal as quadrupedal.

In the case of Hypsilophodon a number of specimens are available instead of only one. This makes the problem of its restoration a simpler matter than in Scelidosaurus. Moreover, there is in America a closely allied form, Laosaurus, of which several species are known. A study of the genus Laosaurus, and the restoration of one species given on Pl. LVII, will clear up several points long in doubt.

Huxley and Hulke both shed much light on this interesting genus, Hypsilophodon; indeed, on many of the Dinosauria. The mystery of the dinosaurian pelvis, which baffled Cuvier, Mantell, and Owen, was mainly solved by them, the ilium and ischium by Huxley, and the pubis by Hulke. The more perfect American specimens have demonstrated the correctness of nearly all their conclusions.

IGUANODON.

PLATE LXXXV.

The fourth restoration here given, that of *Iguanodon Bernissartensis* Boulenger, 1881, one-fortieth natural size, has been made in outline for comparison with American forms. It is based mainly on photographs of the well-known Belgian specimens, the originals of which the writer has studied with considerable care during several visits to Brussels. The descriptions and figures of Dollo[2] have also been used in the preparation of this restoration. A few changes only have been introduced in

---

[1] Philosophical Transactions, 1882.   [2] Bulletin Royal Museum of Belgium, 1882-88.

the accompanying plate, based mainly upon a study of the original specimens.

Besides the four genera here represented, no other European dinosaurs at present known are sufficiently well preserved to admit of accurate restorations of the skeleton. This is true, moreover, of the dinosaurian remains from other parts of the world outside of North America.

AFFINITIES OF DINOSAURS.

The extinct reptiles known as dinosaurs were for a long time regarded as a peculiar order, having, indeed, certain relations to birds, but without being closely allied to any of the groups of known reptiles. Megalosaurus and Iguanodon, the first dinosaurian genera described, were justly considered as representing two distinct families, one including the carnivores, and the other the herbivorous forms.

With the discovery and investigation of Cardiodon (Cetiosaurus) and its allies in Europe, and especially of the gigantic forms with similar characters in America, it became evident that these reptiles could not be placed in the same families with Megalosaurus or Iguanodon, but constituted a well-marked group by themselves. It was this new order, the Sauropoda, as the writer has named them, that first showed definite

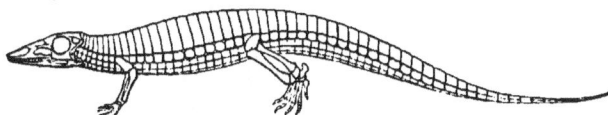

FIG. 56.—Restoration of *Aëtosaurus ferratus* Fraas; with dermal armor of the limbs removed. One-eighth natural size.

characters allying them with other known groups of reptiles. In 1878 he pointed out that the Sauropoda were the least specialized of the dinosaurs, and gave a list of characters in which they showed such an approach to the Mesozoic crocodiles as to suggest a common ancestry at no very remote period.[1]

AFFINITIES WITH AËTOSAURIA.

Again, in 1884, the writer called attention to the same point, and also to the relationship of dinosaurs with the Aëtosauria, as he has named them, a group of small reptiles from the Triassic of Germany showing strong affinities with crocodilians.[2] A restoration of one of these small animals is shown in fig. 56. In the same communication he compared with dinosaurs another allied group, the Hallopoda, which he described from the lower Jurassic of America, but had not then fully investigated. Subsequent researches proved the latter group to be of the first importance in estimating the affinities of dinosaurs, and in figs. 59 and 60 are restorations of the fore and hind limbs of the type species (*Hallopus victor*).

---

[1] American Journal of Science, Vol. XVI. p. 412, November, 1878.
[2] Report British Association, Montreal Meeting, 1884. p. 765.

AFFINITIES WITH BELODONTIA.

Another group of extinct reptiles, which may be termed the Belo-
dontia, were considered in the same paper as allies of the Dinosauria.
They are known from the Trias of Europe and America, and the type
genus, Belodon, has been investigated by many anatomists, who all
appear to have regarded it as a crocodilian, an opinion that in the
light of our present knowledge may fairly be questioned.

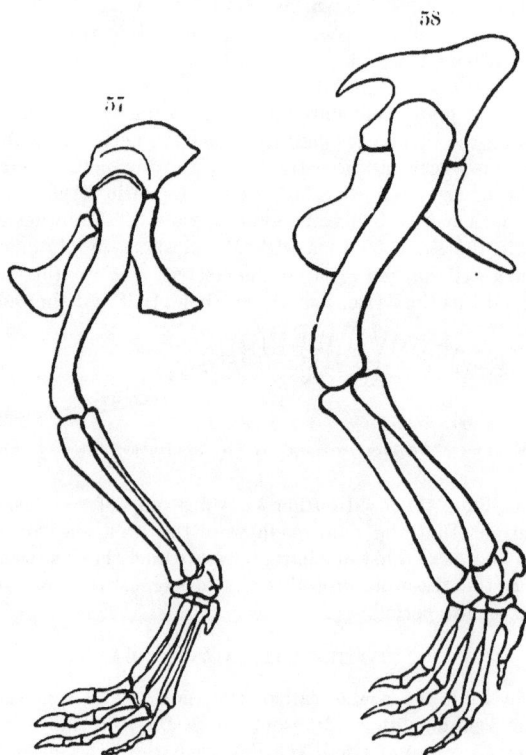

FIG. 57.—Diagram of left hind limb of *Alligator mississippiensis* Gray ; seen from the left ; in position
for comparison with dinosaurs. One-fourth natural size.
FIG. 58.—Diagram of left hind limb of *Aëtosaurus ferratus*; in same position. One-half natural size.

AFFINITIES WITH CROCODILIA.

The relations of these various groups to the true crocodiles on the
one hand and to dinosaurs on the other is much too broad a subject to
be introduced here, but attention may at least be called to some points
of resemblance between the dinosaurs and these supposed crocodilian
forms that seem to indicate genetic affinities.

If some of the characteristic parts of the skeletons of these groups
are compared. e. g., of the true Crocodilia as existing to-day, the Belo-
dontia, the Aëtosauria, and the Hallopoda, and all with the corresponding portions of the more typical dinosaurs, the result may indicate in
some measure the relationship between them. Taking first the pelvis
and hind limb, as being especially characteristic. it will be seen in

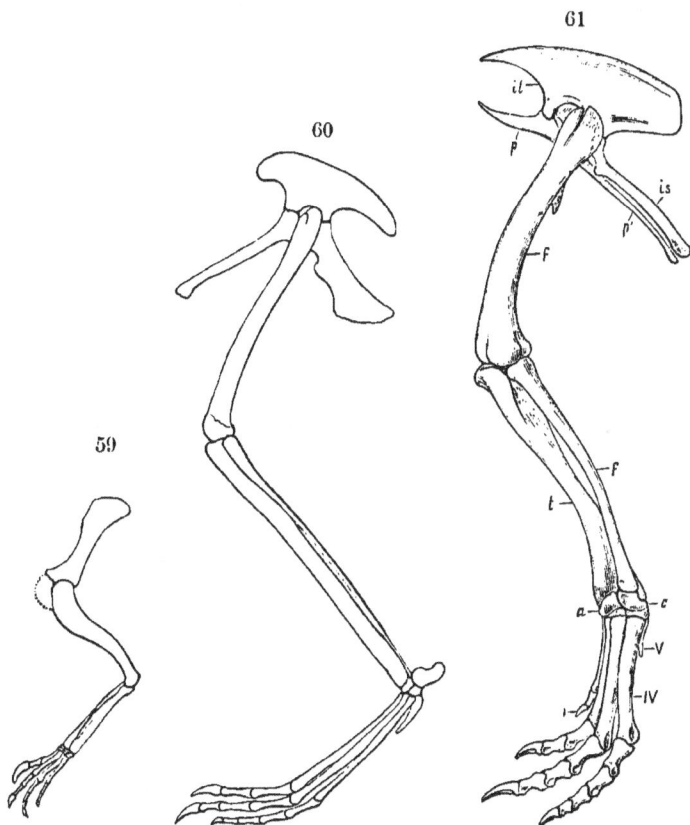

FIG. 59.—Diagram of left fore limb of *Hallopus victor* Marsh; seen from the left.
FIG. 60.—Diagram of left hind limb of same individual. Both figures are one-half natural size.
FIG. 61.—Left hind leg of *Laosaurus consors* Marsh; outside view. One-sixth natural size.
a, astragalus; *c*, calcaneum; *f*, femur; *f'*, fibula; *il*, ilium· *is*, ischium; *p*, pubis; *p'*, postpubis;
t, tibia; *I, IV, V*, first, fourth, and fifth digits.

the existing alligator, as represented in fig. 57, that the pubic bone is
excluded from the acetabulum, articulating with the ischium only, and
not at all with the ilium. The calcaneum, moreover, has a posterior
extension. In Aëtosaurus, as shown in fig. 58, the pubic bone forms
part of the acetabulum, as in dinosaurs and birds, and this is a note-

worthy difference from all the existing crocodiles. · The hind foot, however, is of the crocodilian type, with the calcaneum showing a posterior
projection.

In Belodon, only the pelvis of which is here represented (fig. 62),
the pubis contributes a very important part to the formation of the
acetabulum, and to the entire pelvic arch. The latter differs from the
pelvis of a typical dinosaur mainly in the absence of an open acetabulum, but a moderate enlargement of the fontanelle at the junction of
the three pelvic elements would practically remove this difference. A
more erect position of the limb, leading to a more distinct head on the
femur, might possibly bring about such a result. The feet and limbs
of Belodon are crocodilian in type.

Bearing these facts in mind, the diagram representing the restored
fore and hind limbs of the diminutive Hallopus (figs. 59–60) shows first

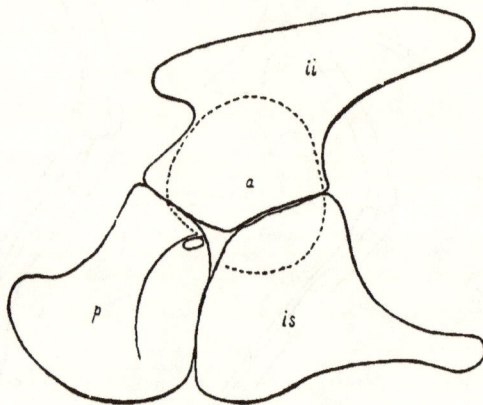

FIG. 62.—Diagram of pelvis of *Belodon Kapffi* von Meyer; seen from the left. One-fourth natural
size.

a, acetabular surface within dotted line; *il*, ilium; *is*, ischium; *p*, pubis.

of all the true dinosaurian pelvis, with the pubic bone taking part
in the open acetabulum, and forming an important and distinctive
element of the pelvic arch. The delicate posterior limb and foot,
evidently adapted mainly for leaping, as the generic name suggests,
are quite unique among the Reptilia, but the tarsus, especially the
calcaneum, recalls strongly the same region in the orders already
passed in review.

Just what this posterior extension of the calcaneum signifies in this
case it is difficult to decide from the evidence now known. It may be
merely an adaptive character, as Hallopus appears in nearly every
other respect to be a true carnivorous dinosaur. It may, however, be
an inheritance from a crocodilian ancestry, preserved by a peculiar
mode of life. Whatever its origin may have been, it was certainly,

during the life of the animal, an essential part of the remarkable leaping foot to which it belonged, and in which it has since kept its position undisturbed. The presence of such an element in the foot of this
diminutive dinosaur certainly suggests that the group Hallopoda,
which the writer has here considered a suborder, stands somewhat apart
from the typical Theropoda, but not far enough away to be excluded
from the subclass Dinosauria, as defined in the present paper.

The genus Plateosaurus (Zanclodon), which is from essentially the

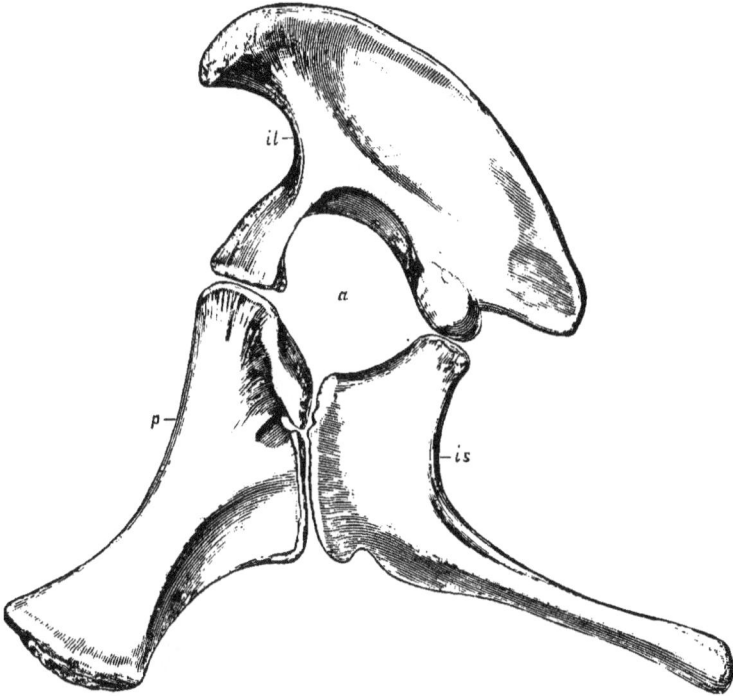

FIG. 63.—Pelvis of *Morosaurus lentus* Marsh; seen from the left. One-eighth natural size.
a, acetabular opening; other letters as in fig. 62.

same geological horizon in Germany as Aëtosaurus and Belodon, is one
of the oldest true dinosaurs known, and a typical member of the order
Theropoda. In the pelvic arch of this reptile the ilium and ischium are
in type quite characteristic of the group to which it belongs, but the
pubic elements are unique. They consist of a pair of broad, thin
plates united together so as to form an apron-like shield in front, quite
unlike anything known in other dinosaurs. The wide pubic bones of
Belodon, and the corresponding plates in some of the Sauropoda (Morosaurus, fig. 63), indicate that this feature of the reptilian pelvis may

have been derived from some common ancestor of a generalized primitive type. The known transformations of this same pelvic element in one other order of dinosaurs (the Predentata) show that the modifications here suggested are well within the limits of probability. The hind limb of one genus of this order is shown in fig. 61 (p. 233).

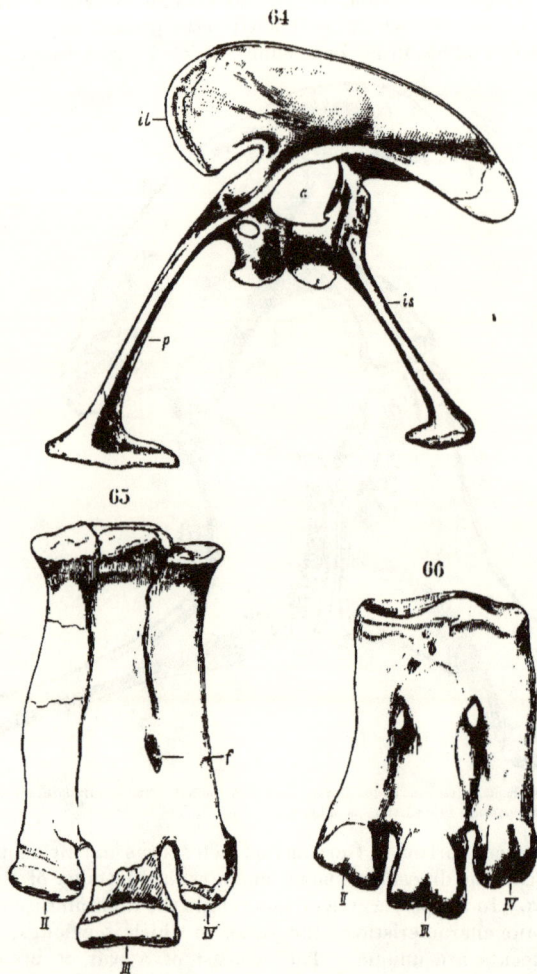

FIG. 64.—Pelvis of *Ceratosaurus nasicornis* Marsh; seen from the left. One-twelfth natural size. Letters as in fig. 63.

FIG. 65.—United metatarsal bones of *Ceratosaurus nasicornis*; left foot; front view. One-fourth natural size.

FIG. 66.—United metatarsal bones of great penguin (*Aptenodytes Pennantii* G. R. Gr.); left foot, front view. Natural size.

The skulls of Aëtosaurus and Belodon both show features characteristic of some of the dinosaurs, especially of the Sauropoda, but these
features need not be discussed here.

AFFINITIES WITH BIRDS.

The relation of dinosaurs to birds, a subject of importance, must
also be postponed for another occasion. One point, however, may be
mentioned in this connection. The pelvic bones of all known birds,
living and extinct, except the genus Archæopteryx, are coossified,
while in all the known dinosaurs they are separate, excepting Ceratosaurus (fig. 64) and Ornithomimus. Again, all known adult birds, living and extinct, with possibly the single exception of Archæopteryx,
have the metatarsal bones firmly united (fig. 66), while all the Dinosauria, except Ceratosaurus (fig. 65), have these bones separate. The
exception in each case brings the two classes near together at this point,
and their close affinity is thus rendered more than probable.

These few facts will throw some light on the affinities of the reptiles
known as the Dinosauria. The problem is certainly one of much difficulty, and the writer hopes soon to discuss it more fully elsewhere.

PART V.

CLASSIFICATION OF DINOSAURIA.

In the present review of the dinosaurs the writer has confined himself mainly to the type specimens which he has described, but has
included with them other important remains where these were available
for investigation. The extensive collections in the museum of Yale
University contain so many of the important type specimens now known
from America that they alone furnish an admirable basis for classification, and it was mainly upon these that he first established the present system, which has since been found to hold equally good for the
dinosaurs discovered elsewhere. In the further study of these reptiles
it was also necessary to examine both the European forms and those
from other parts of the world, and he has now studied nearly every
known specimen of importance. These investigations have enabled
him to make this classification more complete, and to bring it down to
the present time.

Many attempts have been made to classify the dinosaurs, the first
being that of Hermann von Meyer in 1830. The name Dinosauria,
proposed for the group by Owen in 1839, has been generally accepted,
although not without opposition. Hæckel, Cope, and Huxley followed,
the last in 1869 proposing the name Ornithoscelida for the order, and
giving an admirable synopsis of what was then known of these strange
reptiles and their affinities. Since then, Hulke, Seeley, Lydekker,
Gaudry, Dollo, Baur, and others have added much to our knowledge
of these interesting animals.

The remarkable discoveries in North America, however, have changed the whole subject, and in place of fragmentary specimens many entire skeletons of dinosaurian reptiles have been brought to light, and thus definite information has replaced uncertainty and rendered a comprehensive classification for the first time possible.

The system of classification first proposed by the writer in 1881 has been very generally approved, but a few modifications have been suggested by others that will doubtless be adopted. This will hardly be the case with several radical changes recently advocated, based mainly upon certain theories of the origin of dinosaurs. At present these theories are not supported by a sufficient number of facts to entitle them to the serious consideration of those who have made a careful study of these reptiles, especially the wonderful variety of forms recently made known from America.

Further discoveries may in time solve the problem of the origin of all the reptiles now called dinosaurs, but the arguments hitherto advanced against their being a natural group are far from conclusive. The idea that the Dinosauria belong to two or more distinct groups, each of independent origin, can at present only claim equal probability with a similar suggestion recently made in regard to mammals. This subject of the origin of the dinosaurs and the relation of their divisions to each other will be more fully treated by the writer elsewhere.

A classification of any series of extinct animals is of necessity, as the writer has previously said, merely a temporary convenience, like the bookshelves in a library, for the arrangement of present knowledge. In view of this fact and of the very limited information in regard to so many dinosaurs known only from fragmentary remains, it will suffice for the present, or until further evidence is forthcoming, to still consider the Dinosauria as a subclass of the great group of Reptilia.

Regarding, then, the dinosaurs as a subclass of the Reptilia, the forms best known at present may be classified as follows:

Subclass DINOSAURIA Owen.

Premaxillary bones separate; upper and lower temporal arches; no teeth on palate; rami of lower jaw united in front by cartilage only. Neural arches of vertebræ joined to centra by suture; cervical and thoracic ribs double-headed; ribs without uncinate processes; sacral vertebræ united; caudal vertebræ numerous; chevrons articulated intervertebrally. Scapula elongate; no precoracoid; clavicles wanting. Ilium prolonged in front of the acetabulum; acetabulum formed in part by pubis; ischia meet distally on median line. Fore and hind limbs present, the latter ambulatory, and larger than those in front; head of femur at right angles to condyles; tibia with procnemial crest; fibula complete; first row of tarsals composed of astragalus and calcaneum only, which together form the upper portion of ankle joint; reduction in number of digits begins with the fifth.

Order THEROPODA (Beast foot). Carnivorous.

Skull with external narial openings lateral; large antorbital vacuity; brain case incompletely ossified; no pineal foramen; premaxillaries with teeth; no predentary bone; dentary without coronoid process; teeth with smooth compressed crowns and crenulated edges. Vertebræ more or less cavernous; posterior trunk vertebræ united by diplosphenal articulation. Neural canal in sacrum of moderate size. Each sacral rib supported by two vertebræ; diapophyses distinct from sacral ribs. Sternum unossified. Pubes projecting downward, and united distally; no postpubis. Fore limbs small; limb bones hollow; astragalus closely applied to tibia; feet digitigrade; digits with prehensile claws; locomotion mainly bipedal.

(1) Family Megalosauridæ. Lower jaws with teeth in front. Anterior vertebræ convexo-concave; remaining vertebræ biconcave; five sacral vertebræ. Abdominal ribs. Ilium expanded in front of acetabulum; pubes slender, and distally coossified. Femur longer than tibia; astragalus with ascending process; five digits in manus and four in pes.

Genus Megalosaurus (Poikilopleuron). Jurassic and Cretaceous. Known forms, European.

(2) Family Dryptosauridæ. Lower jaws with teeth in front. Cervical vertebræ opisthocœlian; remaining vertebræ biconcave; sacral vertebræ less than five. Ilium expanded in front of acetabulum; distal ends of pubes coossified and much expanded; an interpubic bone. Femur longer than tibia; astragalus with ascending process; fore limbs very small, with compressed prehensile claws. (Pls. X-XII.)

Genera Dryptosaurus (Lælaps), Allosaurus, Cœlosaurus, Creosaurus. Jurassic and Cretaceous. All from North America.

(3) Family Labrosauridæ. Lower jaws edentulous in front. Cervical and dorsal vertebræ convexo-concave; centra cavernous or hollow. Pubes robust, with anterior margins united; an interpubic bone. Femur longer than tibia; astragalus with ascending process. (Pl. XIII.)

Genus Labrosaurus. Jurassic, North America.

(4) Family Plateosauridæ (Zanclodontidæ). Vertebræ biconcave; two sacral vertebræ. Ilium expanded behind acetabulum; pubes broad, elongate plates, with anterior margins united; no interpubic bone; ischia united at distal ends. Femur longer than tibia; astragalus without ascending process; five digits in manus and pes.

Genera Plateosaurus (Zanclodon), Teratosaurus (?), Dimodosaurus. Triassic. Known forms, European.

(5) Family Anchisauridæ. Skull light in structure, with recurved, cutting teeth. Vertebræ plane or biconcave. Bones hollow. Ilium expanded behind acetabulum; pubes rod-like and not coossified distally; no interpubic bone. Fore limbs well developed; femur longer than tibia; astragalus without ascending process; five digits in manus and in pes. (Pls. II-IV.)

Genera Anchisaurus (Megadactylus), Ammosaurus, Arctosaurus (?),

Bathygnathus, and Clepsysaurus, in North America; and in Europe, Palæosaurus, Thecodontosaurus. All known forms, Triassic.

Suborder CŒLURIA (Hollow tail).

(6) Family Cœluridæ. Teeth much compressed. Vertebræ and bones of skeleton very hollow or pneumatic; neural canal much expanded; anterior cervical vertebræ convexo-concave; remaining vertebræ biconcave; anterior cervical ribs coossified with vertebræ; pubes slender and distally coossified; an interpubic bone. Femur shorter than tibia; metatarsals very long and slender. (Pl. VII.)

Genera Cœlurus in North America, and Aristosuchus in Europe. Jurassic.

Suborder COMPSOGNATHA.

(7) Family Compsognathidæ. Skull elongate, with slender jaws and pointed teeth. Cervical vertebræ convexo-concave, with free ribs; remaining vertebræ biconcave. Ischia with long symphysis on median line. Bones very hollow; femur shorter than tibia; astragalus with long ascending process; three functional digits in manus and in pes. (Pl. LXXXII.)

Genus Compsognathus. Jurassic. Only known specimen, European.

Suborder CERATOSAURIA (Horned saurians).

(8) Family Ceratosauridæ. Horn on skull; teeth large and trenchant. Cervical vertebræ plano-concave; remaining vertebræ biconcave. Ribs free. Pelvic bones coossified; ilium expanded in front of acetabulum; pubes slender; an interpubic bone; sacral vertebræ five; ischia slender, with distal ends coossified. Limb bones hollow; manus with four digits; femur longer than tibia; astragalus with ascending process; metatarsals coossified; three digits only in pes. Osseous dermal plates. (Pls. VIII-X, XIV.)

Genus Ceratosaurus. Jurassic, North America.

(9) Family Ornithomimidæ. Pelvic bones coossified with each other and with sacrum; ilium expanded in front of acetabulum. Limb bones very hollow; fore limbs very small; digits with very long, pointed claws; hind limbs of true avian type; femur longer than tibia; astragalus with long ascending process; feet with three functional digits, digitigrade and unguiculate. (Pl. LVIII.)

Genus Ornithomimus. Cretaceous, North America.

Suborder HALLOPODA (Leaping foot).

(10) Family Hallopidæ. Vertebræ and limb bones hollow; vertebræ biconcave; two vertebræ in sacrum. Acetabulum formed by ilium, pubis, and ischium; pubes rod-like, projecting downward, but not coossified distally; no postpubis; ischia with distal ends expanded, meeting below on median line. Fore limbs very small, with four digits in

manus; femur shorter than tibia; hind limbs very long, with three functional digits in pes, and metatarsals greatly elongated; astragalus without ascending process; calcaneum much produced backward; feet digitigrade, unguiculate. (Pl. VI.)

Genus Hallopus. Jurassic, North America.

Order SAUROPODA (Lizard foot). Herbivorous.

External nares at apex of skull; premaxillary bones with teeth; teeth with rugose crowns more or less spoon-shaped; large antorbital openings; no pineal foramen; alisphenoid bones; brain case ossified; no columellæ; postoccipital bones; no predentary bone; dentary without coronoid process. Cervical ribs coossified with vertebræ; anterior vertebræ opisthocœlian, with neural spines bifid; posterior trunk vertebræ united by diplosphenal articulation; presacral vertebræ hollow; each sacral vertebra supports it own transverse process, or sacral rib; no diapophyses on sacral vertebræ; neural canal much expanded in sacrum; first caudal vertebræ procœlian. Sternal bones parial; sternal ribs ossified. Ilium expanded in front of acetabulum; pubes projecting in front, and united distally by cartilage; no postpubis. Limb bones solid; fore and hind limbs nearly equal; metacarpals longer than metatarsals; femur longer than tibia; astragalus not fitted to end of tibia; feet plantigrade, ungulate; five digits in manus and pes; second row of carpal and tarsal bones unossified; locomotion quadrupedal.

(1) Family Atlantosauridæ. A pituitary canal; large fossa for nasal gland. Distal end of scapula not expanded. Sacrum hollow; ischia directed downward, with expanded extremities meeting on median line. Anterior caudal vertebræ with lateral cavities; remaining caudals solid. (Pls. XV–XXIV, and XLII.)

Genera Atlantosaurus, Apatosaurus, Barosaurus, Brontosaurus. Include the largest known land animals. Jurassic, North America.

(2) Family Diplodocidæ. External nares superior; no depression for nasal gland; two antorbital openings; large pituitary fossa; dentition weak, and in front of jaws only; brain inclined backward; dentary bone narrow in front. Ischia with shaft not expanded distally, directed downward and backward, with sides meeting on median lines. Sacrum hollow, with three vertebræ. Caudal vertebræ deeply excavated below; chevrons with both anterior and posterior branches. (Pls. XXV–XXIX.)

Genus Diplodocus. Jurassic, North America.

(3) Family Morosauridæ. External nares lateral; large fossa for nasal gland; small pituitary fossa; dentary bone massive in front; teeth very large. Shaft of scapula expanded at distal end. Sacral vertebræ four in number, and nearly solid; ischia slender, with twisted shaft directed backward, and sides meeting on median line. Anterior caudals solid. (Pls. XXIX–XXXVIII.)

Genera Morosaurus, Camarasaurus (?) (Amphicœlias). Jurassic, North America.

16 GEOL, PT 1——16

(4) Family Pleurocœlidæ. Dentition weak; teeth resembling those of Diplodocus. Cervical vertebræ elongated; centrum hollow, with large lateral openings; sacral vertebræ solid, with lateral depressions in centra; caudal vertebræ solid; anterior caudals with flat articular faces and transversely compressed neural spines; middle caudal vertebræ with neural arch on front half of centrum. Ischia with compressed distal ends, meeting on median line. (Pls. XL and XLI.)

Genera Pleurocœlus, Astrodon (?). Potomac, North America.

(5) Family Titanosauridæ. Fore limbs elongate; coracoid quadrilateral. Presacral vertebræ opisthocœlian; first caudal vertebra biconvex; remaining caudals procœlian; chevrons open above.

Genera Titanosaurus and Argyrosaurus. Cretaceous (?), India and Patagonia.

(6) Family Cardiodontidæ. Teeth of moderate size. Upper end of scapula expanded; humerus elongate; fore limbs near equaling hind limbs in length. Sacrum solid; ischia with wide distal ends meeting on median line. Caudal vertebræ biconcave.

Genera Cardiodon (Cetiosaurus), Bothriospondylus, Ornithopsis, and Pelorosaurus. European, and probably all Jurassic.[1]

Order PREDENTATA. Herbivorous.

Narial opening lateral; no antorbital foramen; brain case ossified; supraorbital bones; teeth with sculptured crowns; maxillary teeth with crowns grooved on outside; lower teeth with grooves on inside of crown; a predentary bone; dentary with coronoid process. Cervical ribs articulating with vertebræ; each sacral rib supported by two vertebræ. Ilium elongated in front of acetabulum; prepubic bones free in front; postpubic bones present; ischia slender, directed backward, with distal ends meeting side to side. Astragalus without ascending process.

Suborder STEGOSAURIA (Plated lizard).

Skull without horns; no teeth in premaxillaries; teeth with distinct compressed crowns and serrated edges. Vertebræ and limb bones solid. Pubes projecting free in front; postpubis present. Fore limbs small; femur longer than tibia; feet plantigrade, ungulate; five digits in manus and four in pes; second row of carpals and tarsals unossified; locomotion mainly quadrupedal. Osseous dermal armor.

(1) Family Stegosauridæ. Vertebræ biconcave. Neural canal in sacrum expanded into large chamber; ischia directed backward, with sides meeting on median line. Dorsal ribs T-shaped in cross section. Astragalus coossified with tibia; metapodials very short. Back surmounted by a crest of vertical plates; tail armed with large spines. (Pls. XLIII–LII.)

Genera Stegosaurus (Hypsirhophus), Diracodon, Palæoscincus, Priconodon, all from North America; and in Europe, Omosaurus, Owen. Jurassic and Cretaceous.

---

[1] The Wealden is here regarded as upper Jurassic, and not Cretaceous. See American Journal of Science, Vol. L, p. 412, November, 1895.

(2) Family Scelidosauridæ. Neural canal narrow; diapophyses of dorsal vertebræ supporting head and tubercle of ribs. Astragalus not coossified with tibia; metatarsals elongated; three functional digits in pes.

Genera Scelidosaurus, Acanthopholis, Hylæosaurus, Polacanthus. Jurassic and Cretaceous. Known forms, all European. (Pl. LXXXIII.)

(3) Family Nodosauridæ. Heavy dermal armor. Bones solid. Fore limbs large; five digits in manus; feet ungulate.

Genus Nodosaurus. Cretaceous, North America.

Suborder CERATOPSIA (Horned face).

Premaxillaries edentulous; teeth with two distinct roots; skull surmounted by massive horn cores; a rostral bone, forming a sharp, cutting beak; expanded parietal crest, with marginal armature; a pineal foramen (?). Vertebræ and limb bones solid; fore limbs large; femur longer than tibia; feet ungulate; locomotion quadrupedal. Dermal armor.

(4) Family Ceratopsidæ. Anterior cervical vertebræ coossified with each other; posterior dorsal vertebræ supporting on the diapophysis both the head and tubercle of the rib; lumbar vertebræ wanting; sacral vertebræ with both diapophyses and ribs. Neural canal in sacrum without marked enlargement. Pubes projecting in front, with distal end expanded; postpubic bone rudimentary or wanting. (Pls. LIX–LXXI.)

Genera Ceratops, Agathaumas, Monoclonius, Polyonax, Sterrholophus, Torosaurus, Triceratops, in North America; and in Europe, Struthiosaurus (Crataeomus). All are Cretaceous.

Suborder ORNITHOPODA (Bird foot).

Skull without horns; premaxillaries edentulous in front. Vertebræ solid. Fore limbs small. Pubes projecting free in front; postpubis present. Astragalus closely fitting to end of tibia; feet digitigrade; three to five functional digits in manus and three to four in pes; locomotion mainly bipedal. No dermal armor.

(5) Family Camptosauridæ (Camptonotidæ). Premaxillaries edentulous; teeth in single row; a supraorbital fossa. Anterior vertebræ opisthocœlian; sacral vertebræ five, not coossified, with peg-and-notch articulation. Sternum unossified. Limb bones hollow; fore limbs small; five digits in manus. Postpubis reaching to the distal end of ischium. Femur longer than tibia, and with pendent fourth trochanter; hind feet with three functional digits. (Pls. LIII–LVI.)

Genus Camptosaurus (Camptonotus). Jurassic, North America.

(6) Family Laosauridæ. Premaxillaries edentulous; teeth in single row. Anterior vertebræ with plane articular faces; sacral vertebræ coossified. Sternum unossified. Postpubis reaching to distal end of ischium. Limb and foot bones hollow; fore limbs very small; five digits in manus; femur shorter than tibia; metatarsals elongate; four digits in pes. (Pls. LV and LVII.)

Genera Laosaurus and Dryosaurus.   Jurassic, North America.

(7) Family Hypsilophodontidæ.   Premaxillaries with teeth; teeth in single row; sclerotic bony plates.   Anterior vertebræ opisthocœlian; sacral vertebræ coossified.   Sternum ossified.   Postpubis extending to end of ischium.   Limb bones hollow; five digits in manus; femur shorter than tibia; hind feet with four digits.   (Pl. LXXXIV.)

Genus Hypsilophodon.   Wealden, England.

(8) Family Iguanodontidæ.   Premaxillaries edentulous; teeth in single row.   Anterior vertebræ opisthocœlian.   Manus with five digits; pollex spine-like.   Sternal bones ossified.   Postpubis incomplete. Femur longer than tibia; three functional digits in pes.   (Pl. LXXXV.)

Genera Iguanodon, Vectisaurus.   Jurassic and Cretaceous.   Known forms, all European.

(9) Family Trachodontidæ (Hadrosauridæ).   Premaxillaries edentulous; teeth in several rows, forming with use a tessellated grinding surface.   Cervical vertebræ opisthocœlian.   Limb bones hollow; fore limbs small; femur longer than tibia.

Genera Trachodon (Hadrosaurus, Diclonius), Cionodon, and Ornithotarsus.   Cretaceous, North America.

(10) Family Claosauridæ.   Premaxillaries edentulous; teeth in several rows, but a single row only in use.   Cervical vertebræ opisthocœlian.   Limb bones solid; fore limbs small.   Sternal bones parial. Postpubis incomplete.   Sacral vertebræ nine.   Femur longer than tibia; feet ungulate; three functional digits in manus and pes.   (Pls. LXXII–LXXIV.)

Genus Claosaurus.   Cretaceous, North America.

(11) Family Nanosauridæ.   Teeth compressed and pointed, and in a single uniform row.   Cervical and dorsal vertebræ short and biconcave; sacral vertebræ three.   Ilium with very short pointed front and narrow posterior end.   Limb bones and others very hollow; fore limbs of moderate size; humerus with strong radial crest; femur curved, and shorter than tibia; fibula pointed below; metatarsals very long and slender.   Anterior caudals short.

Genus Nanosaurus.   Jurassic, North America.   Includes the most diminutive of known dinosaurs.

POSTSCRIPT.

The accompanying plates, as well as the figures in the text, are all from original drawings made to illustrate the writer's investigations on the early vertebrate life of North America.   Many of these illustrations were designed especially for the monographs on dinosaurian reptiles now in preparation for the United States Geological Survey, and are here used, with the approval of the Director, to give the general reader a clear idea of some of the type specimens of one great group of extinct animals that were long the dominant forms of life on this continent.

YALE UNIVERSITY, *June 15, 1895.*

PLATES.

PLATE II.

PLATE II.

TRIASSIC DINOSAURS.—THEROPODA.

ANCHISAURIDÆ.

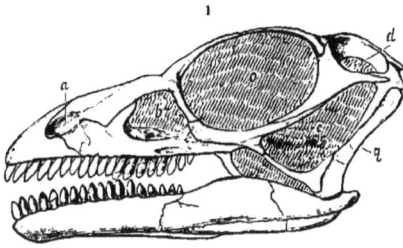

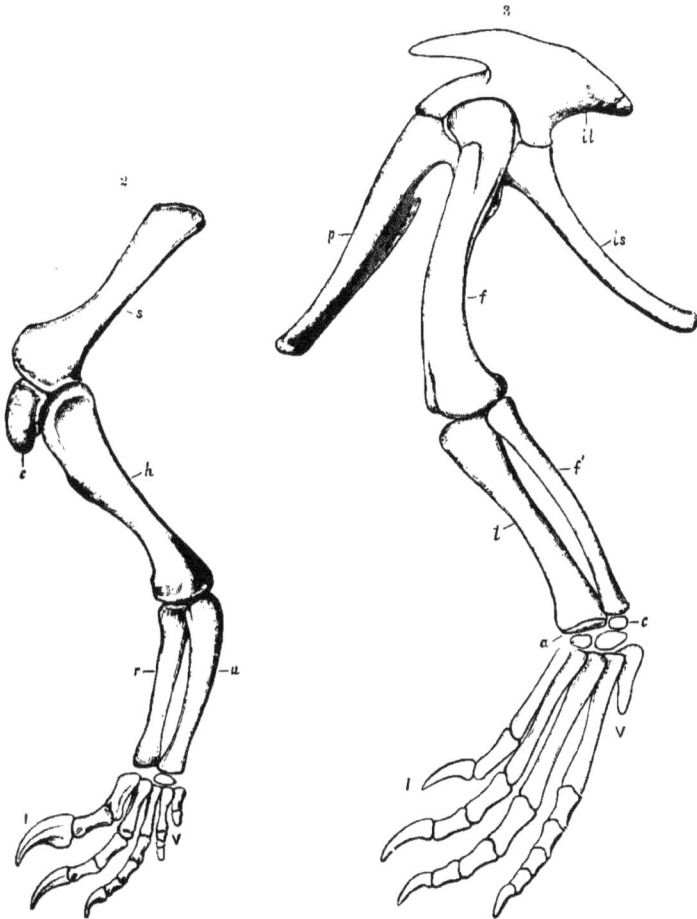

ANCHISAURUS COLURUS Marsh.
Triassic.

PLATE III.

PLATE III.

TRIASSIC DINOSAURS.—THEROPODA.

ANCHISAURIDÆ.

a, nasal opening; *bp,* bas pterygoid process; *d,* upper temporal fossa; *f,* frontal; *j,* jugal; *n,* nasal; *o,* orbit; *oc,* occipital condyle; *p,* parietal; *p',* paroccipital process; *pf,* prefrontal; *pm,* premaxillary.

a, posterior view of distal ends; *ac,* acetabular surface; *il,* face for ilium; *is,* face for ischium; *p,* distal end; *pb,* face for pubis; *s,* symphysis; *1, 2, 3,* sacral vertebræ.

c, centrale; *r.* radiale; *R,* radius; *U,* ulna; *I,* first digit; *V,* fifth digit.

a, astragalus; *c,* calcaneum; *F,* fibula; *T,* tibia; *t2, t3, t4,* tarsal bones; *I,* first digit; *V,* fifth digit.

250

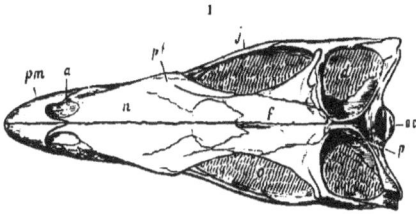

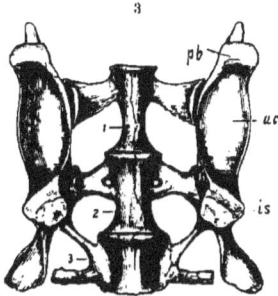

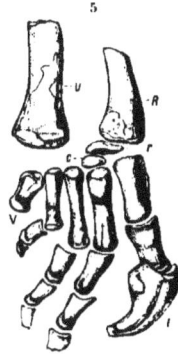

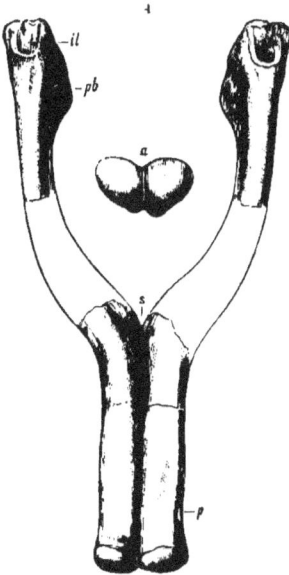

ANCHISAURUS AND AMMOSAURUS.
Triassic.

PLATE IV.

PLATE IV.

Triassic Dinosaurs.—Theropoda.

Anchisauridæ.

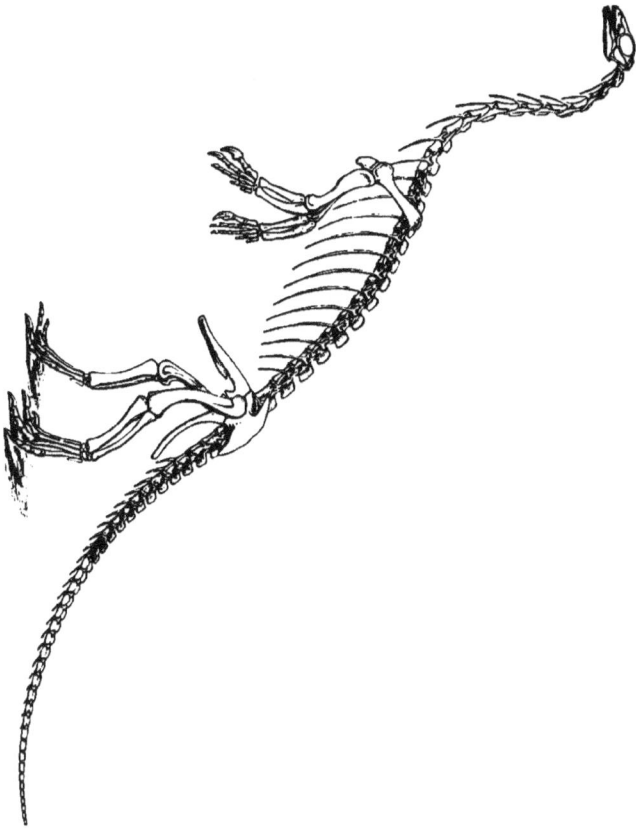

PLATE V.

PLATE V.

TRIASSIC DINOSAURS.—FOOTPRINTS.

The specimens represented in figs. 1–4 are from the Connecticut River sandstone of Massachusetts, and the one shown in fig. 5 is from nearly the same horizon in Arizona.

1

2

3

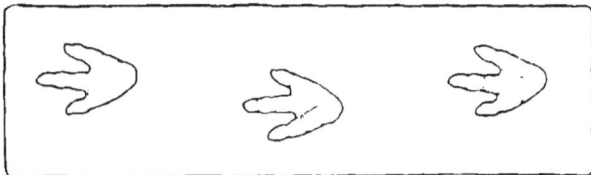

4

5

FOOTPRINTS OF TRIASSIC DINOSAURS.

PLATE VI.

PLATE VI.

Jurassic Dinosaurs.—Theropoda.

Halloptidæ.

256

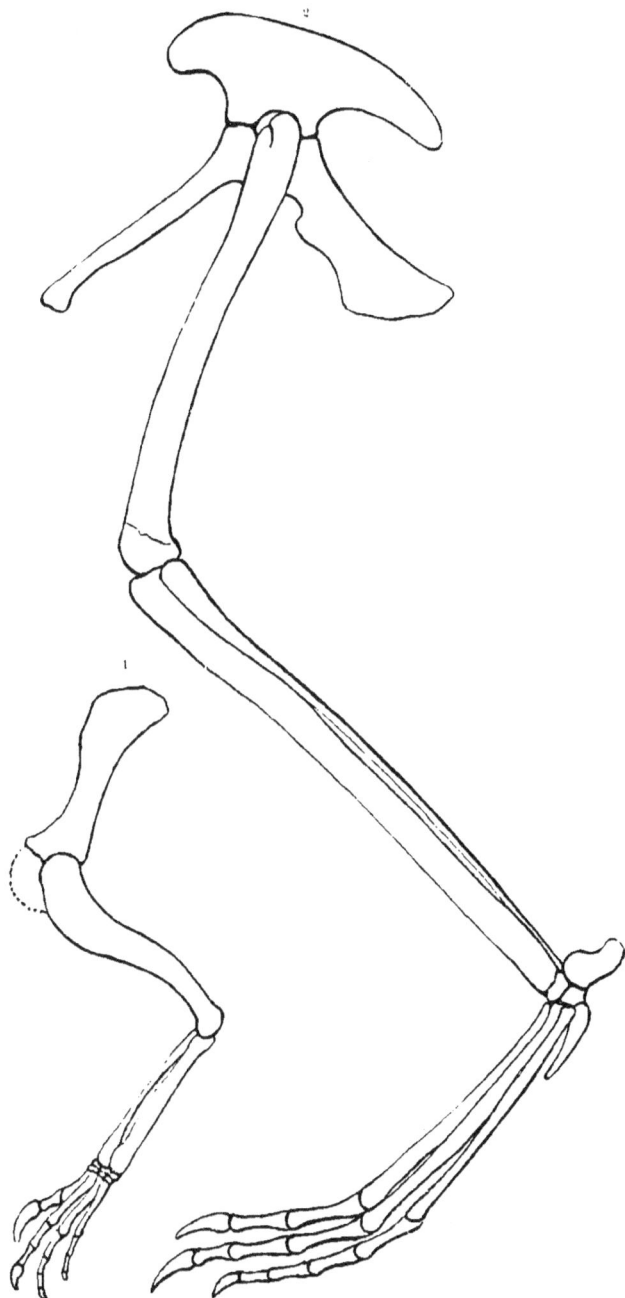

HALLOPUS VICTOR Marsh.
Jurassic.

PLATE VII.

PLATE VII.

JURASSIC DINOSAURS.—THEROPODA.

CŒLURIDÆ.

    a, anterior end; *c*, cavity; *d*, diapophysis; *f*, lateral foramen; *nc*, neural canal; *p*, posterior end; *r*, coossified rib; *s*, neural spine; *z*, anterior zygapophysis; *z'*, posterior zygapophysis.

All the figures of vertebræ are natural size.

258

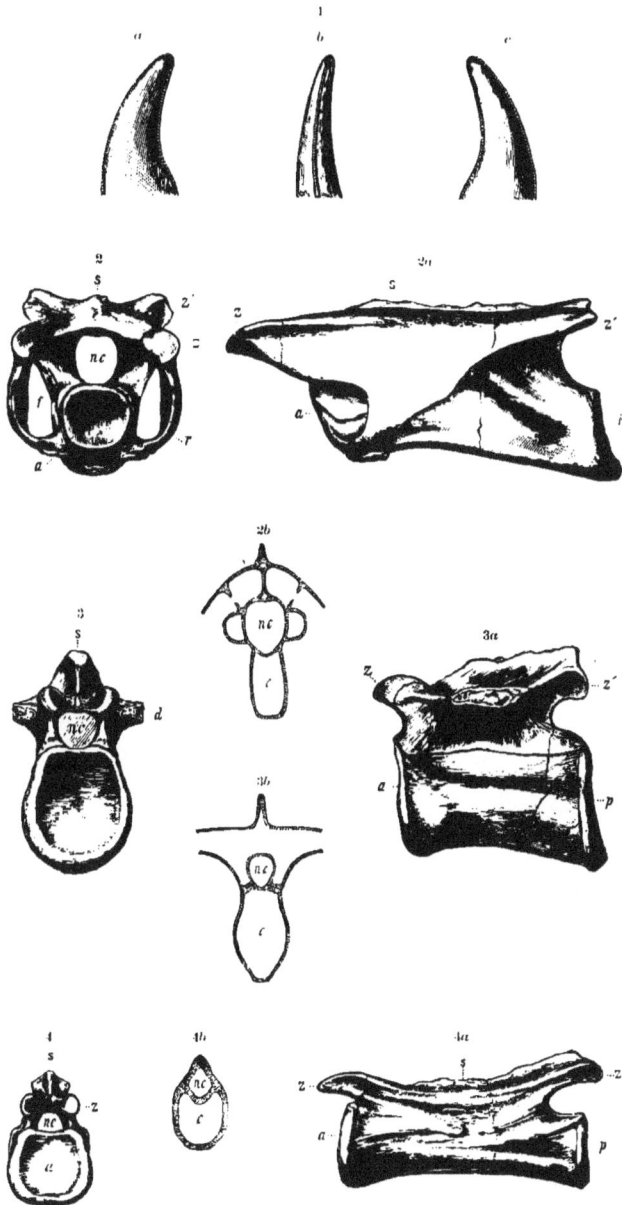

CŒLURUS FRAGILIS Marsh.
Jurassic.

PLATE VIII.

PLATE VIII.

JURASSIC DINOSAURS.—THEROPODA.

CERATOSAURIDÆ.

a, nasal opening; b, horn core; c, antorbital opening; c', cerebral hemispheres; d, orbit; e, lower temporal fossa; f, frontal bone; f', foramen in lower jaw; h, supratemporal fossa; j, jugal; m, maxillary bone; m', medulla; n, nasal bone; oc, occipital condyle; ol, olfactory lobes; pf, prefrontal bone; pm, premaxillary bone; q, quadrate bone; qj, quadratojugal bone; t, transverse bone.

All the figures are one-sixth natural size.

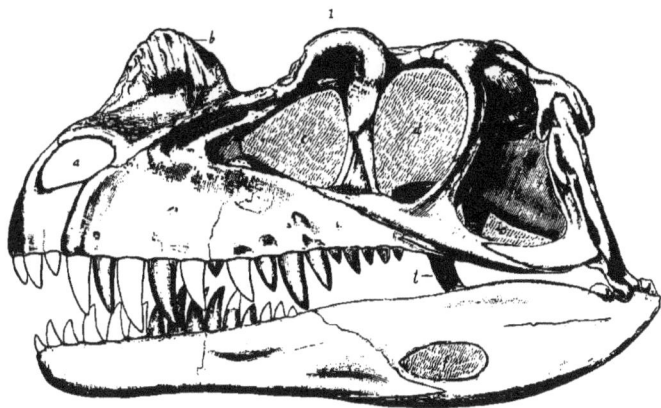

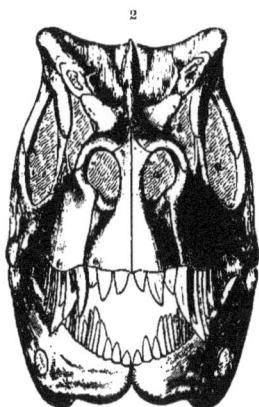

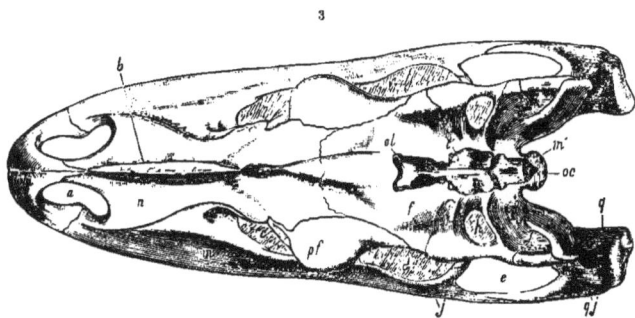

CERATOSAURUS NASICORNIS Marsh.

Jurassic.

PLATE IX.

PLATE IX.

JURASSIC DINOSAURS.—THEROPODA.

CERATOSAURIDÆ.

All the figures are one-sixth natural size.

262

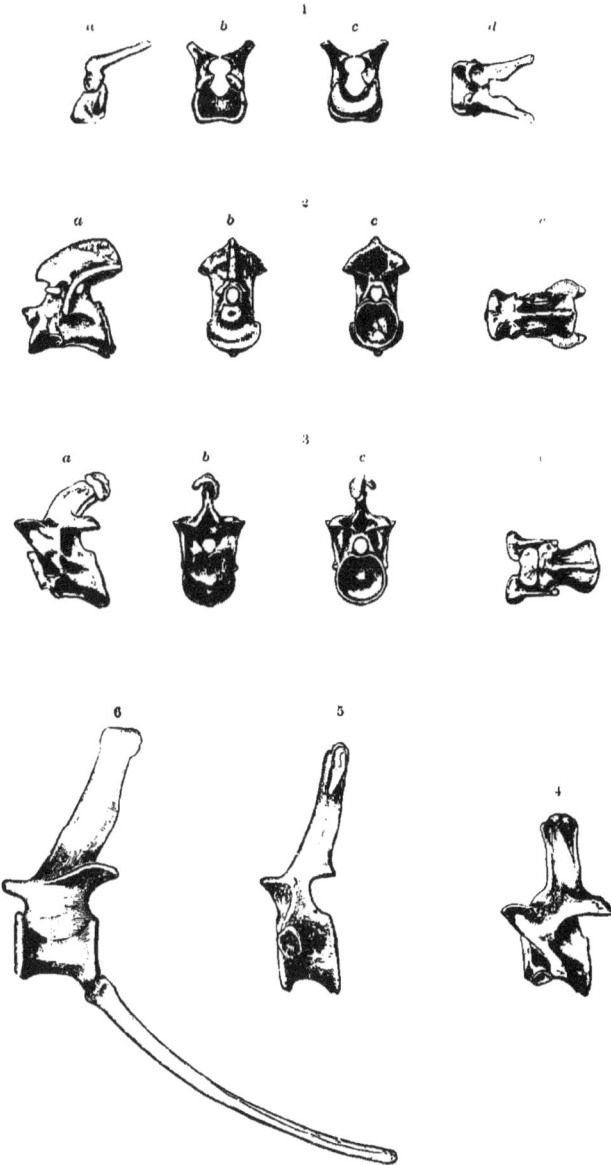

CERATOSAURUS NASICORNIS.

Jurassic.

PLATE X.

PLATE X.

JURASSIC DINOSAURS.—THEROPODA.

CERATOSAURIDÆ, DRYPTOSAURIDÆ, AND COELURIDÆ.

264

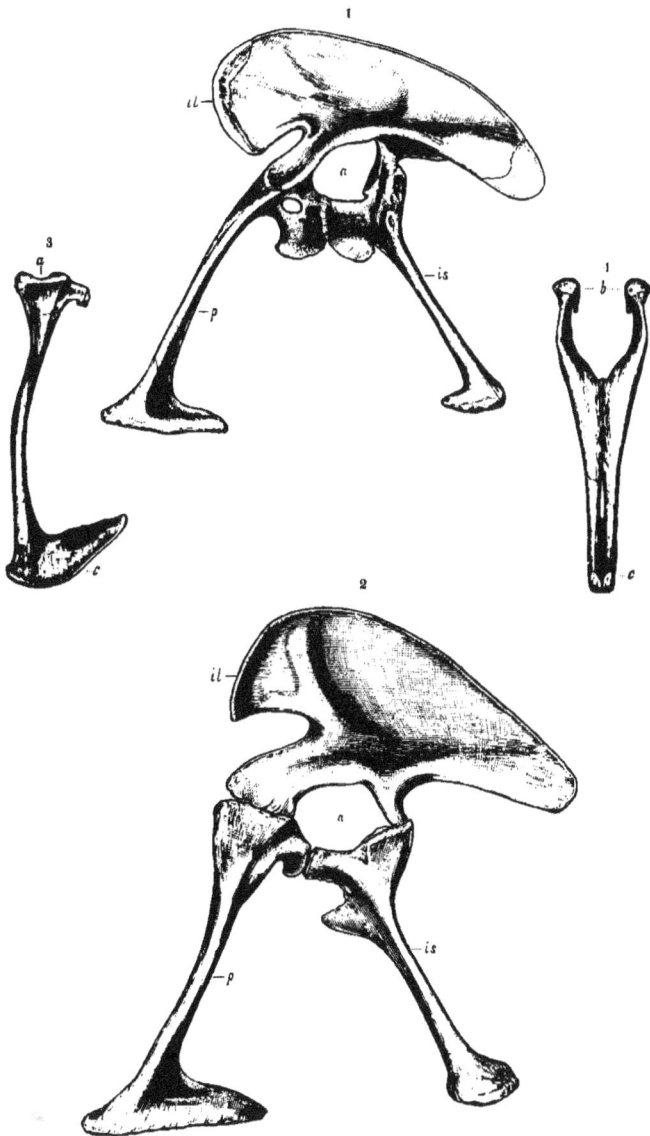

CERATOSAURUS, ALLOSAURUS, AND CŒLURUS.
Jurassic.

PLATE XI.

PLATE XI.

JURASSIC DINOSAURS.—THEROPODA.

DRYPTOSAURIDÆ.

266

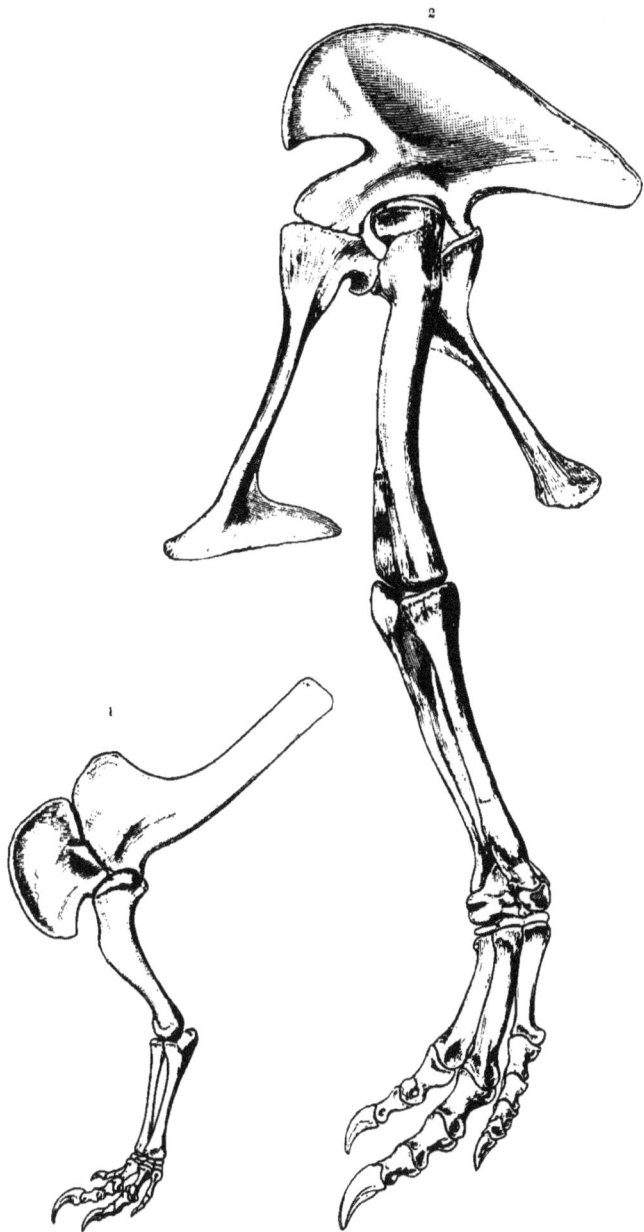

ALLOSAURUS FRAGILIS Marsh.
Jurassic.

PLATE XII.

PLATE XII.

JURASSIC DINOSAURS.—THEROPODA.

DRYPTOSAURIDÆ.

268

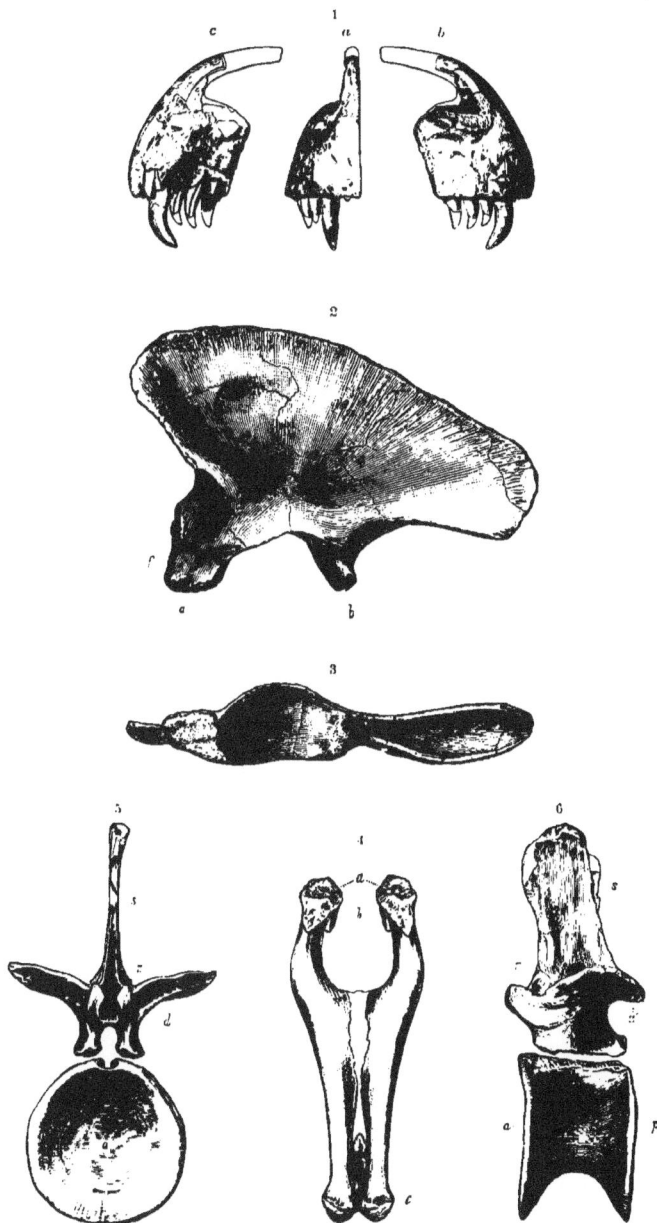

CREOSAURUS ATROX Marsh.
Jurassic.

PLATE XIII.

PLATE XIII.

JURASSIC DINOSAURS.—THEROPODA.

LABROSAURIDÆ.

270

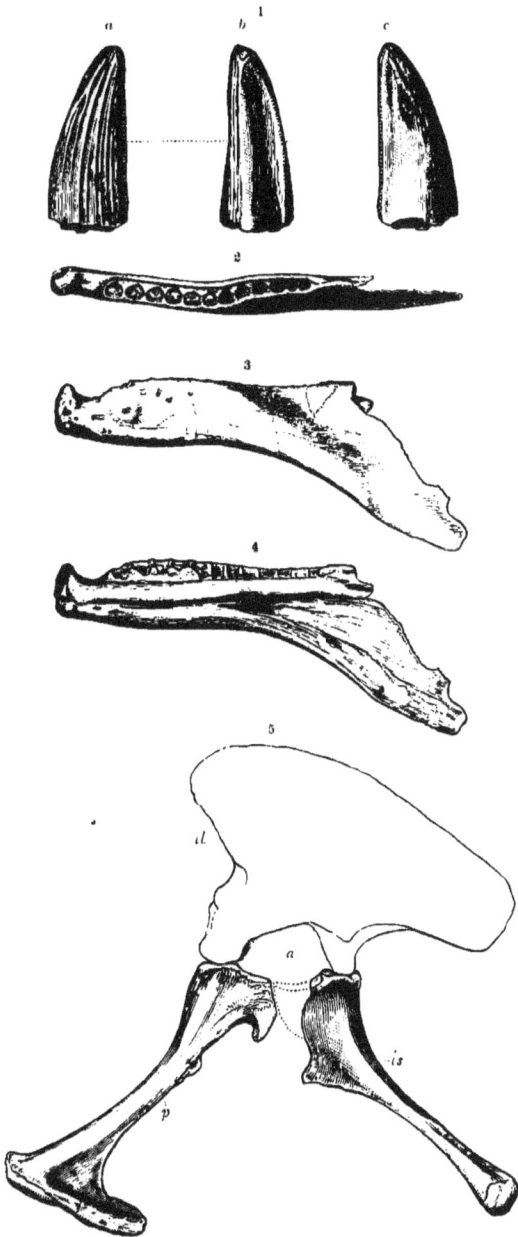

LABROSAURUS.
Jurassic.

PLATE XIV.

PLATE XIV.

Jurassic Dinosaurs.—Theropoda.

Ceratosauridæ.

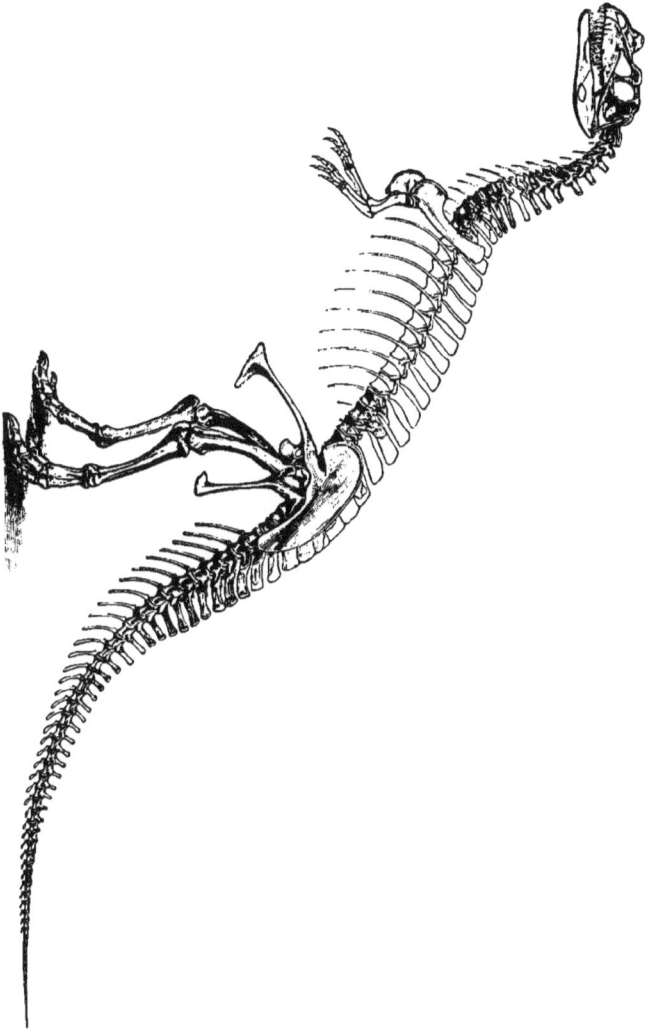

PLATE XV.

PLATE XV.

Jurassic Dinosaurs.—Sauropoda.

Atlantosauridæ.

    bp, basioccipital process; *c'*, pituitary canal; *f*, foramen magnum; *h*, posterior fossa; *i*, internal carotid foramen; *oc*, occipital condyle; *p*, paroccipital process; *pr*, parietal; *s*, suture; *so*, supraoccipital.

    Both figures are one-half natural size.

274

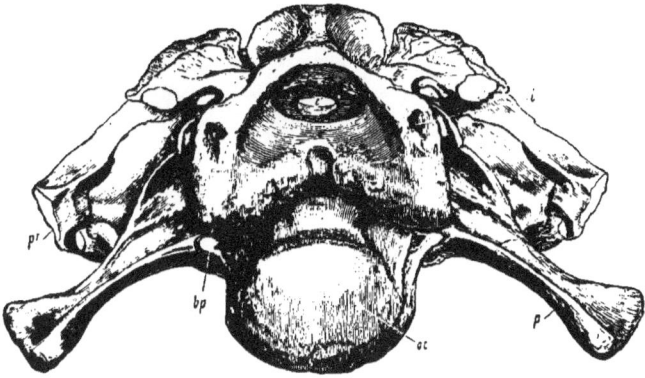

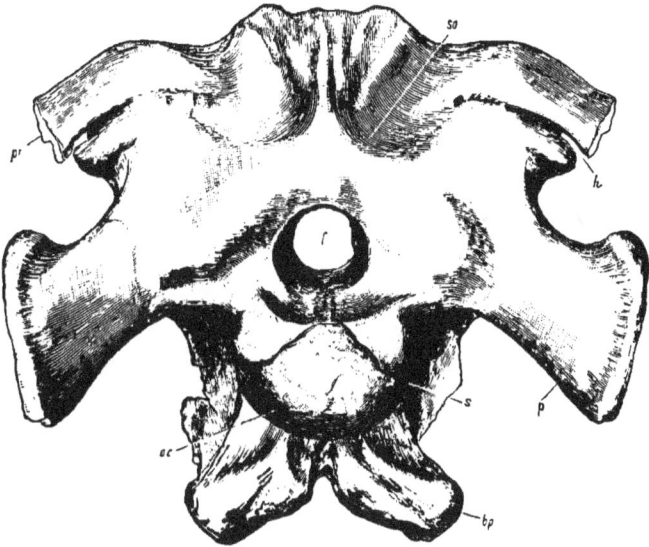

ATLANTOSAURUS MONTANUS Marsh.

Jurassic.

PLATE. XVI.

275

PLATE XVI.

Jurassic Dinosaurs.—Sauropoda.

Atlantosauridæ.

276

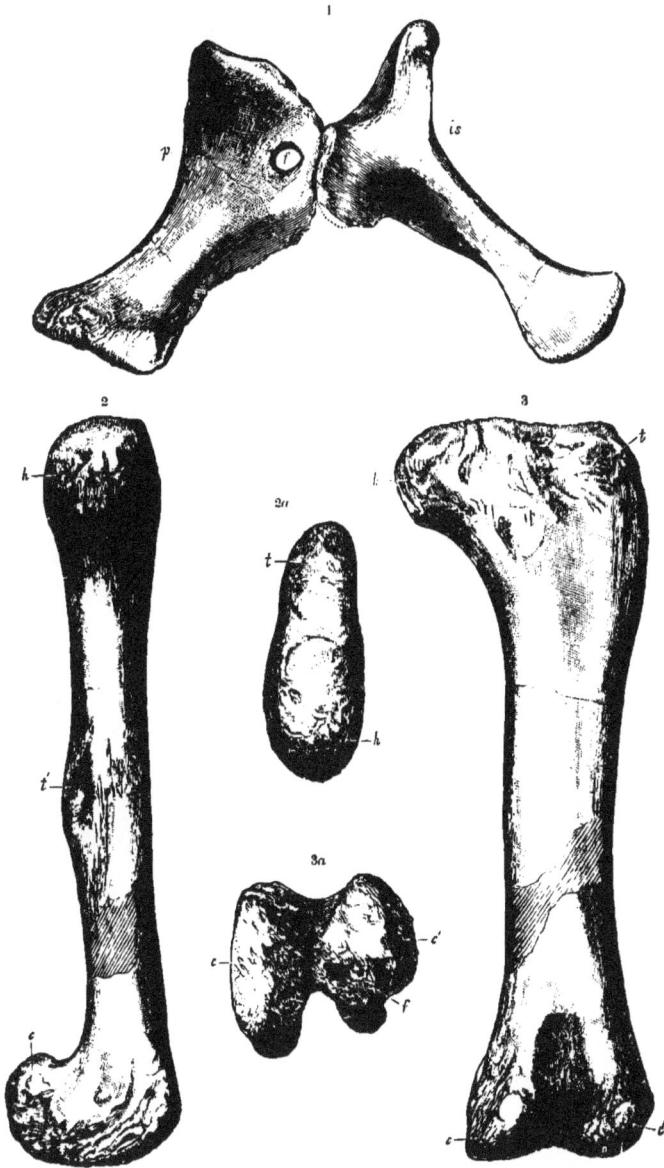

ATLANTOSAURUS IMMANIS Marsh

Jurassic.

PLATE XVII.

PLATE XVII.

Jurassic Dinosaurs.—Sauropoda.

Atlantosauridæ.

a, first sacral vertebra; b, transverse process of first vertebra; c, transverse process of second vertebra; d, transverse process of third vertebra; e, transverse process of fourth vertebra; f, f', f'', foramina between transverse processes; g, surface for union with ilium; p, last sacral vertebra. Both figures are one-tenth natural size.

278

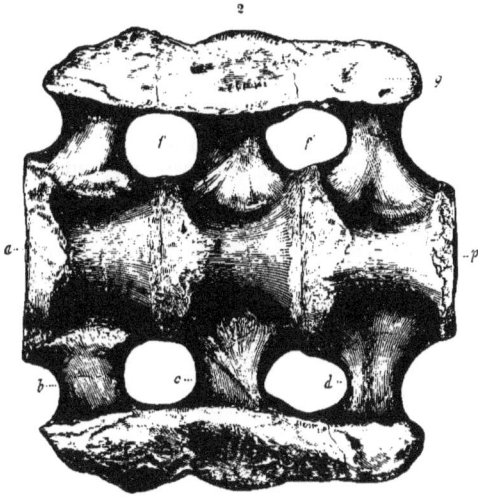

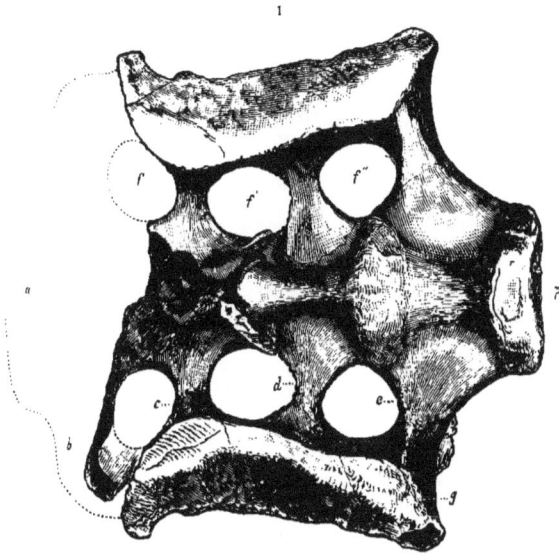

ATLANTOSAURUS AND APATOSAURUS.
Jurassic.

PLATE XVIII.

JURASSIC DINOSAURS.—SAUROPODA.

ATLANTOSAURIDÆ.

280

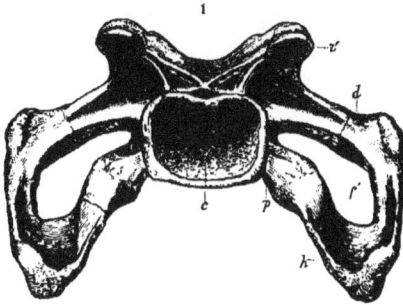

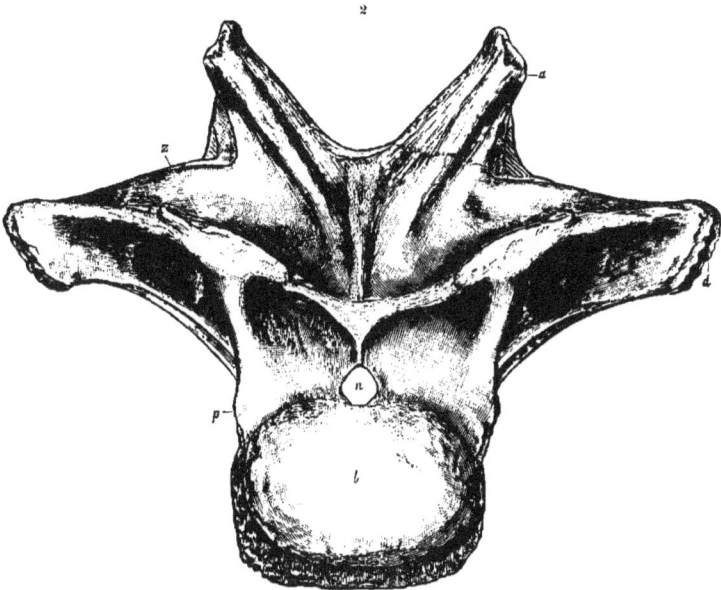

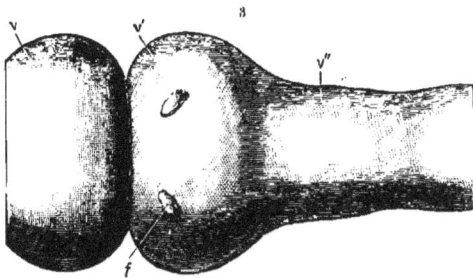

APATOSAURUS.
Jurassic.

PLATE XIX.

PLATE XIX.

JURASSIC DINOSAURS.—SAUROPODA.

ATLANTOSAURIDÆ AND MOROSAURIDÆ.

282

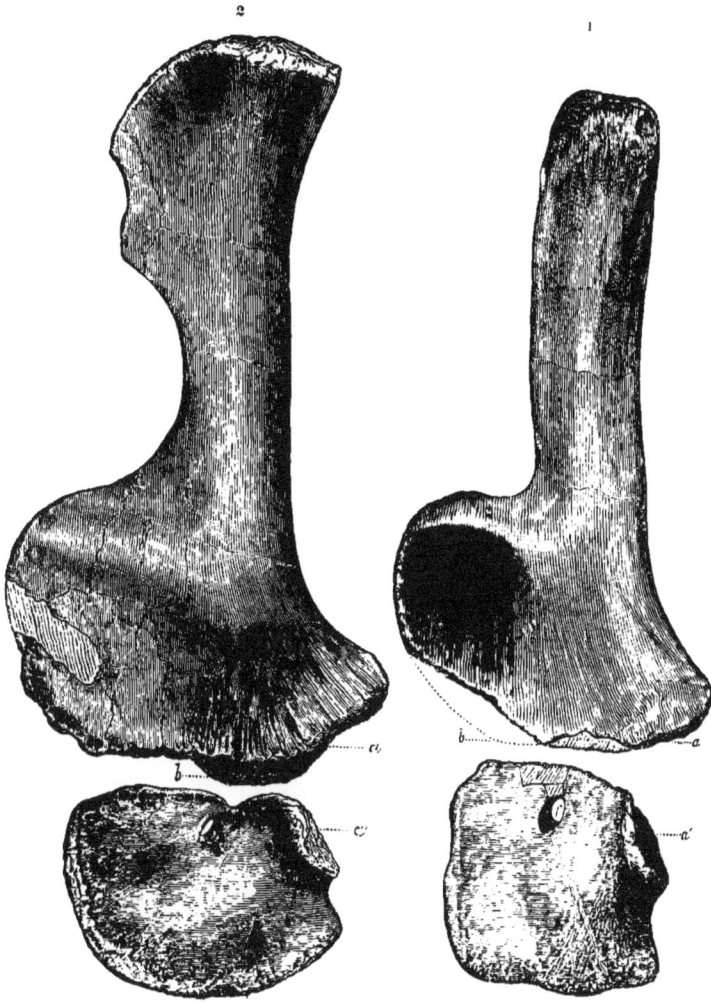

APATOSAURUS AND MOROSAURUS

Jurassic.

PLATE XX.

PLATE XX.

Jurassic Dinosaurs.—Sauropoda.

Atlantosauridæ.

284

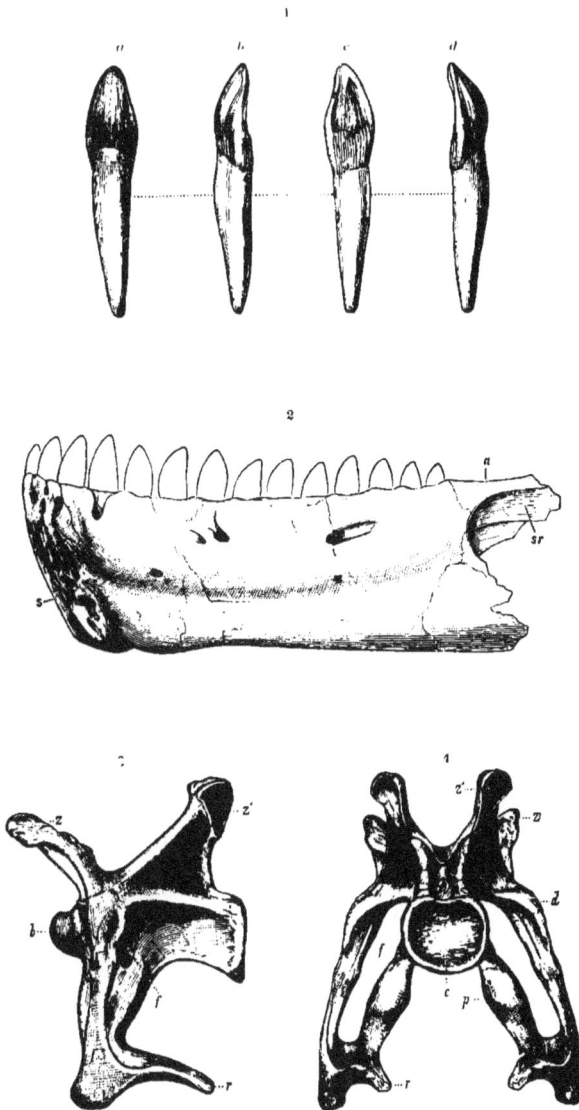

BRONTOSAURUS EXCELSUS Marsh.

Jurassic

PLATE XXI.

PLATE XXI.

Jurassic Dinosaurs.—Sauropoda.

Atlantosauridæ.

b, ball; c, cup; d, diapophysis; f, foramen in centrum; f', lateral foramen; n, neural canal; p, parapophysis; r, rib; s, neural spine; z, anterior zygapophysis; z', posterior zygapophysis.

All the figures are one-twelfth natural size.

286

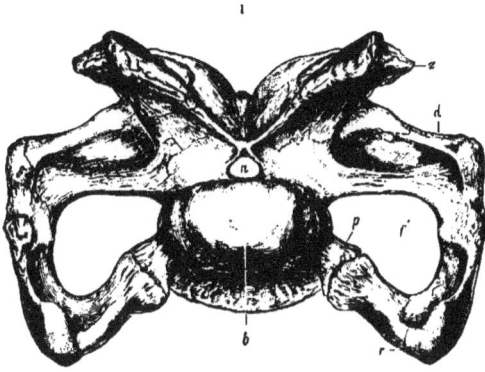

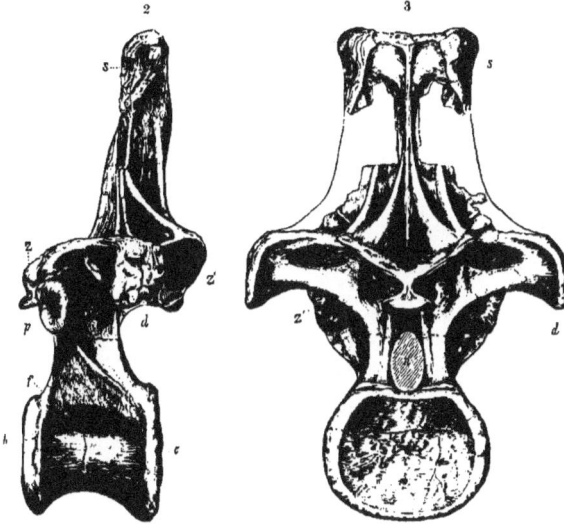

BRONTOSAURUS EXCELSUS
Jurassic.

PLATE XXII.

PLATE XXII.

JURASSIC DINOSAURS.—SAUROPODA.

ATLANTOSAURIDÆ.

288

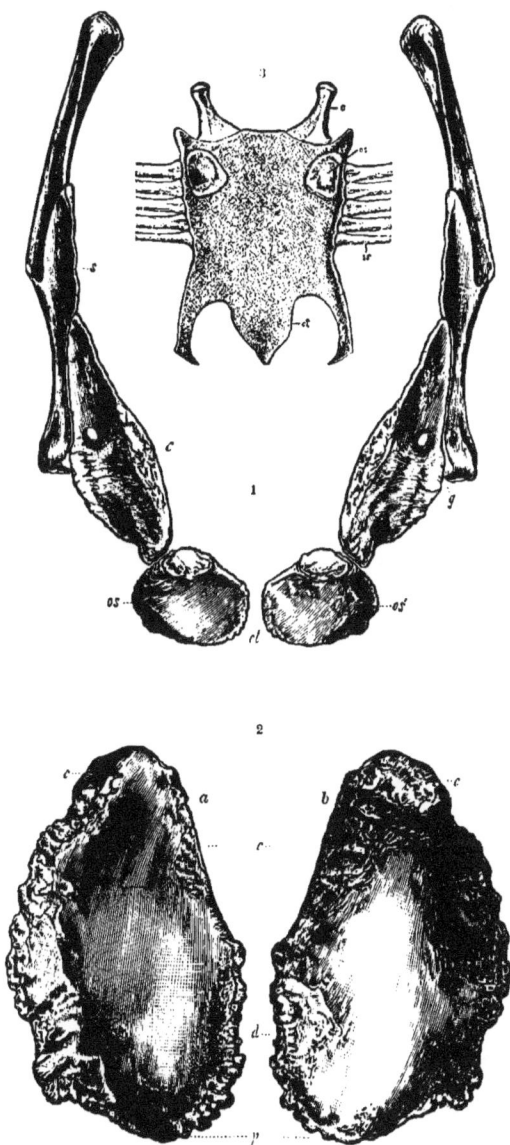

STERNAL PLATES OF BRONTOSAURUS AND YOUNG STRUTHIO.

PLATE XXIII.

PLATE XXIII.

JURASSIC DINOSAURS.—SAUROPODA.

ATLANTOSAURIDÆ.

Page.

FIG. 1. Sacrum of *Brontosaurus excelsus* Marsh; seen from below.............. 170
    a, first sacral vertebra; *b*, transverse process of first vertebra; *c*, transverse
    process of second vertebra; *d*, transverse process of third vertebra; *e*,
    transverse process of fourth vertebra; *f, f', f'', f'''*, foramina between
    processes of sacral vertebræ; *g*, surface for union with ilium; *l*, last
    lumbar vertebra; *p*, last sacral vertebra.

FIG. 2. Section through second vertebra of sacrum of *Brontosaurus excelsus*.... 170
    c, cavity; *g*, face for union with ilium; *nc*, neural canal.
    Both figures are one-tenth natural size.

    290

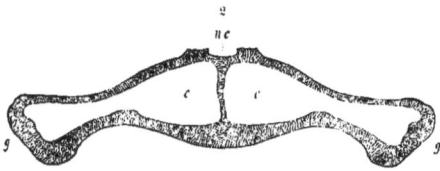

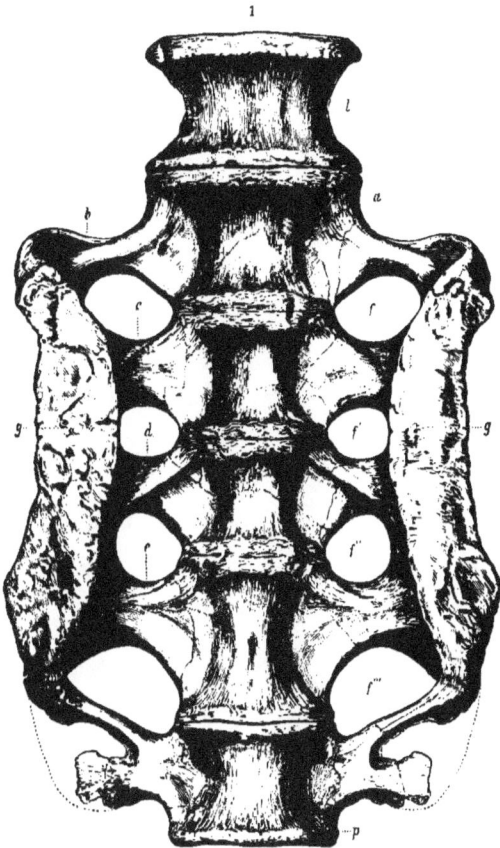

BRONTOSAURUS EXCELSUS.

Jurassic.

PLATE XXIV.

PLATE XXIV.

Jurassic Dinosaurs.—Sauropoda.

Atlantosauridæ.

    292

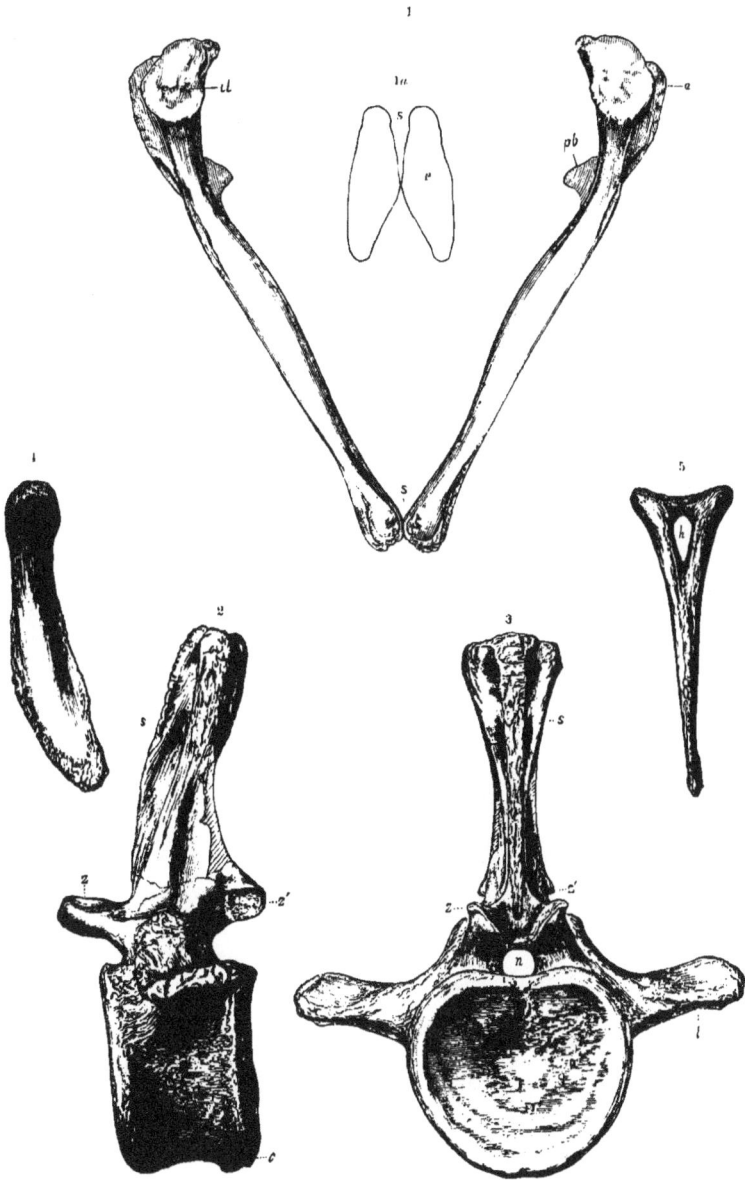

BRONTOSAURUS EXCELSUS.
Jurassic.

PLATE XXV.

PLATE XXV.

Jurassic Dinosaurs.—Sauropoda.

Diplodocidæ.

All the figures are one-sixth natural size.

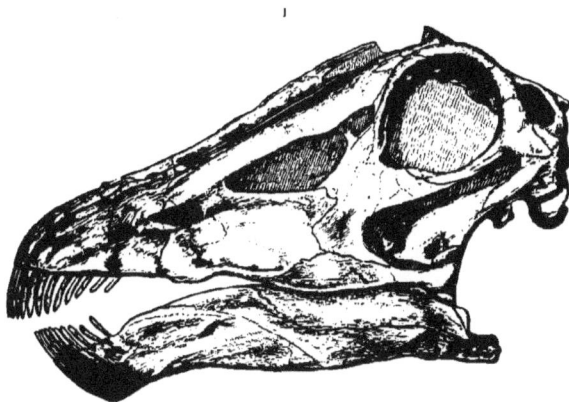

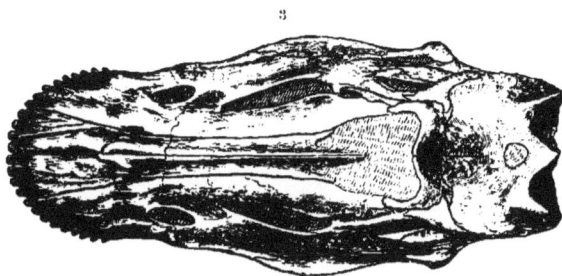

DIPLODOCUS LONGUS Marsh

Jurassic.

PLATE XXVI.

PLATE XXVI.

Jurassic Dinosaurs.—Sauropoda.

Diplodocidæ.

296

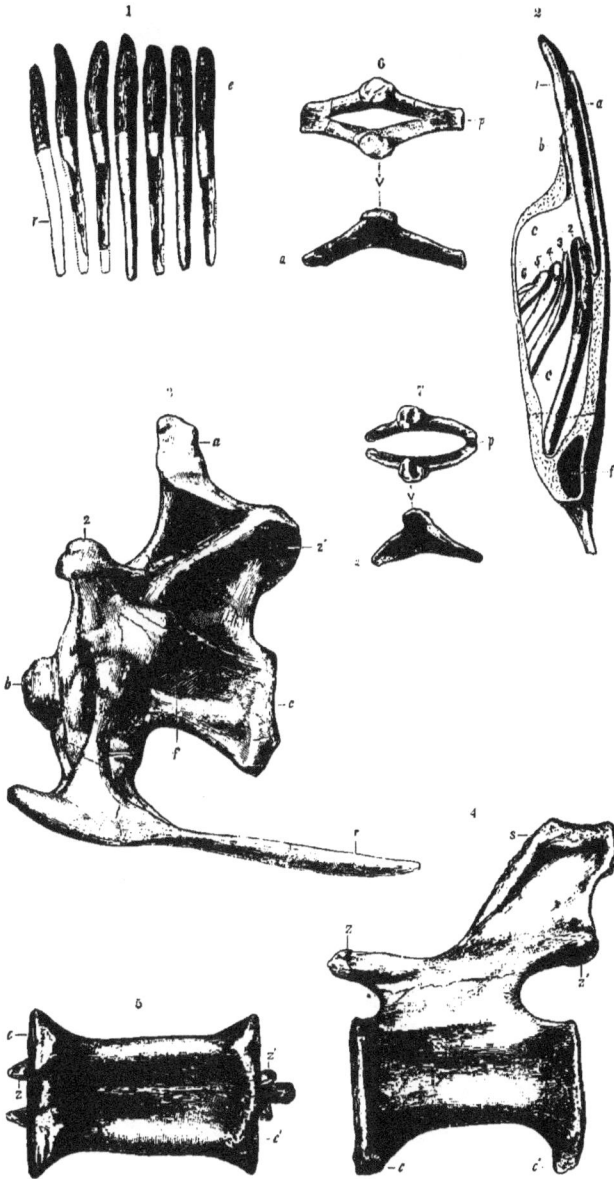

DIPLODOCUS LONGUS
Jurassic.

PLATE XXVII.

PLATE XXVII.

Jurassic Dinosaurs.—Sauropoda.

Diplodocidæ.

298

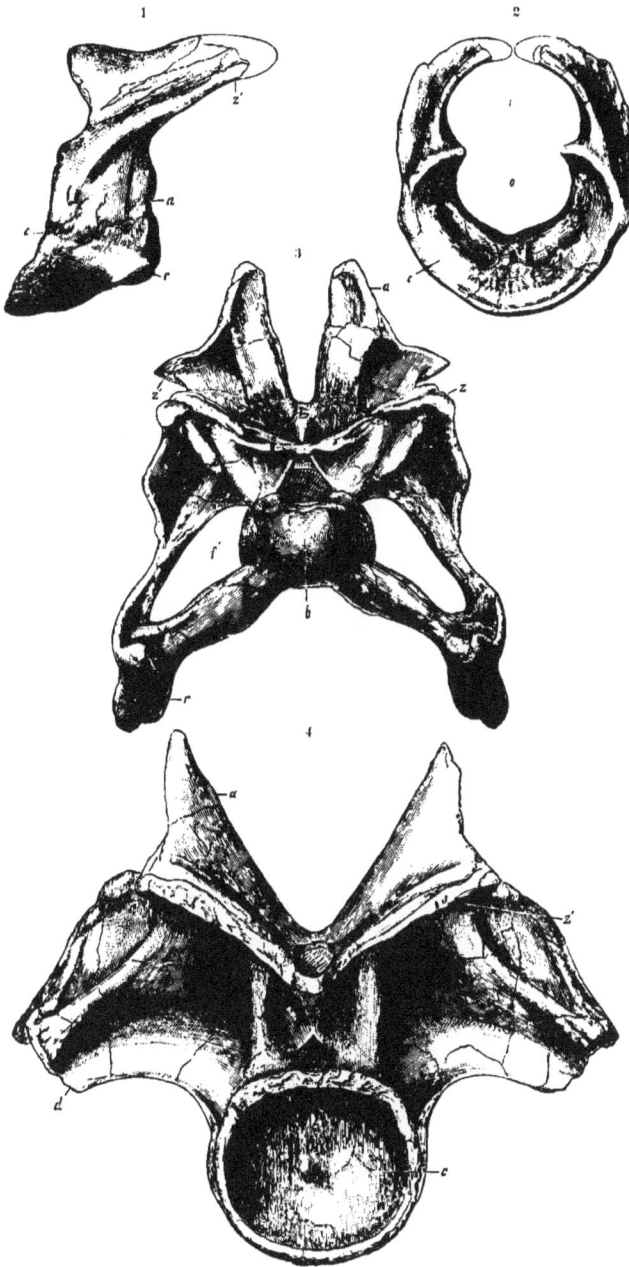

DIPLODOCUS LONGUS

Jurassic.

PLATE XXVIII.

PLATE XXVIII.

Jurassic Dinosaurs.—Sauropoda.

Diplodocidæ.

300

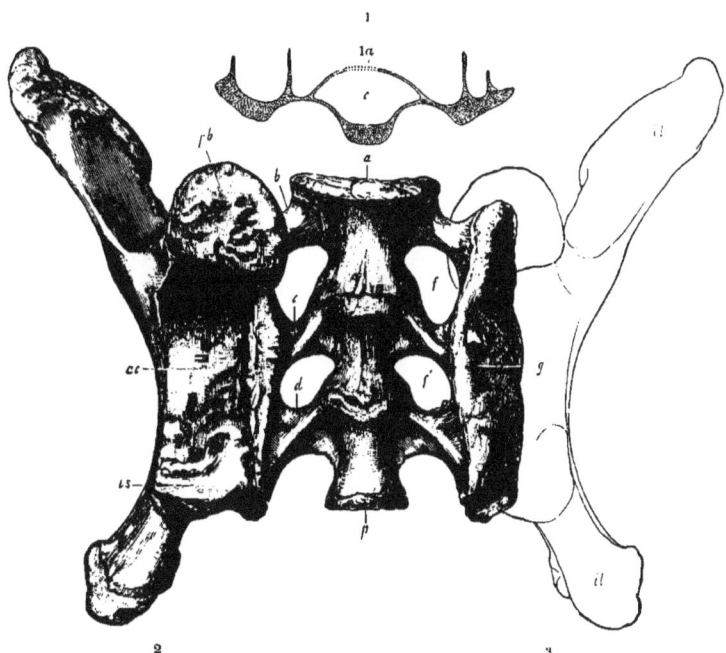

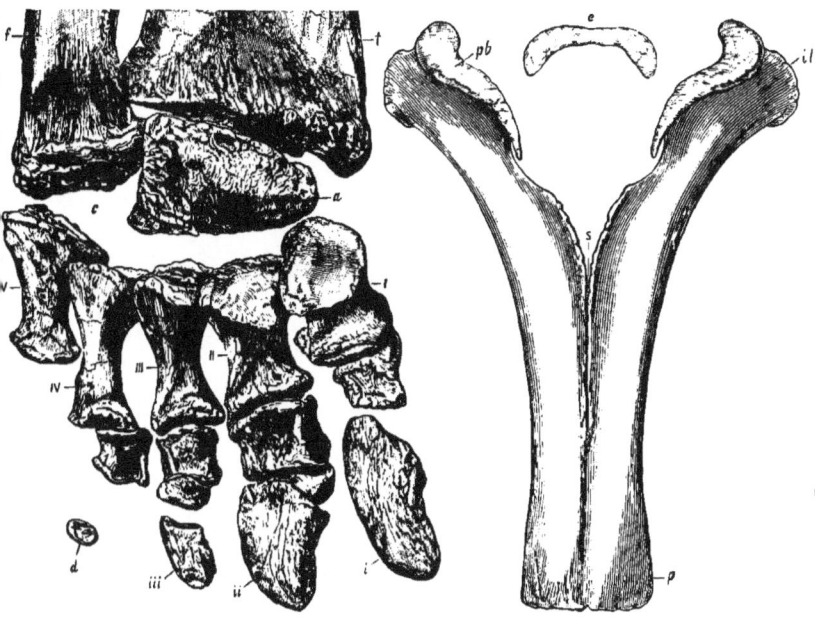

DIPLODOCUS LONGUS.
Jurassic.

PLATE XXIX.

PLATE XXIX.

Jurassic Dinosaurs.—Sauropoda.

Diplodocidæ.

302

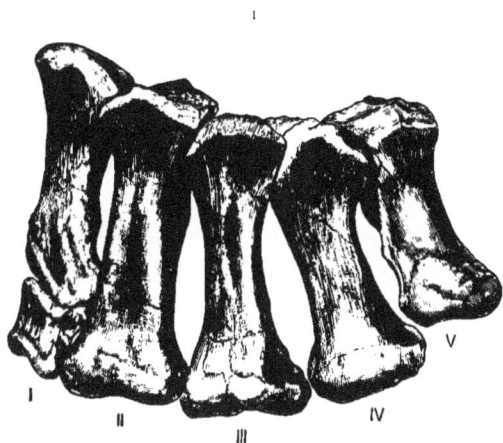

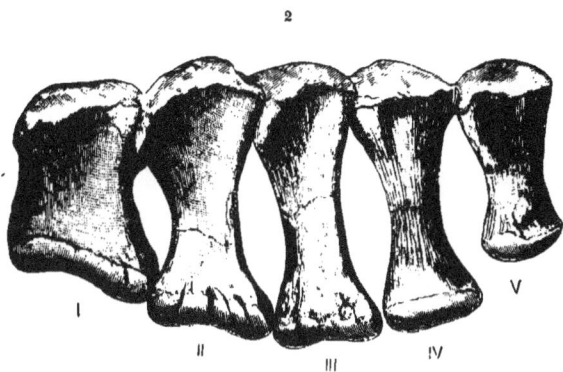

DIPLODOCUS AND MOROSAURUS.
Jurassic

PLATE XXX.

PLATE XXX.

Jurassic Dinosaurs.—Sauropoda.

Morosauridæ.

304

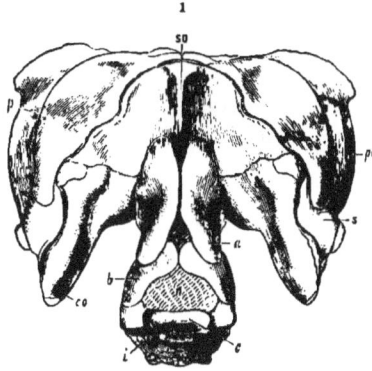

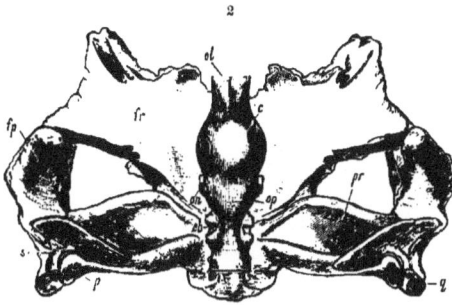

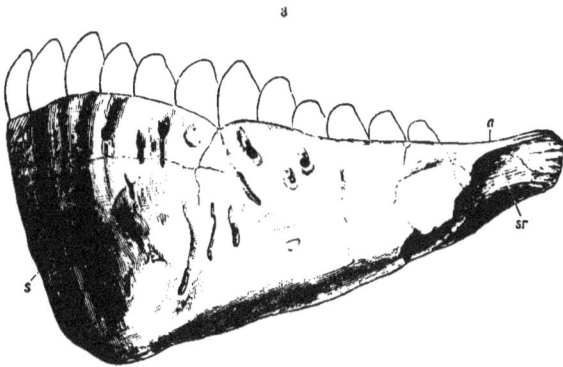

MOROSAURUS.
Jurassic.

PLATE XXXI.

PLATE XXXI

Jurassic Dinosaurs.—Sauropoda.

Morosauridæ.

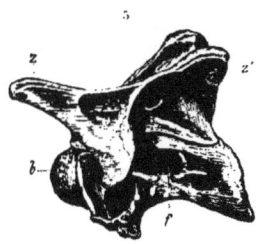

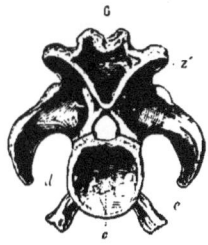

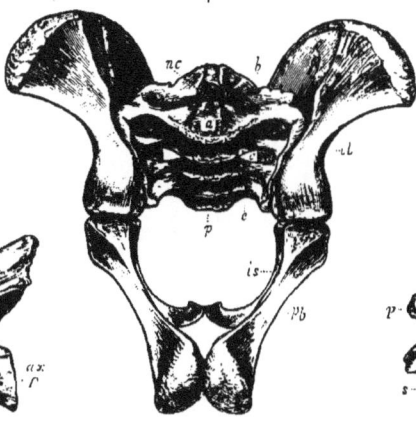

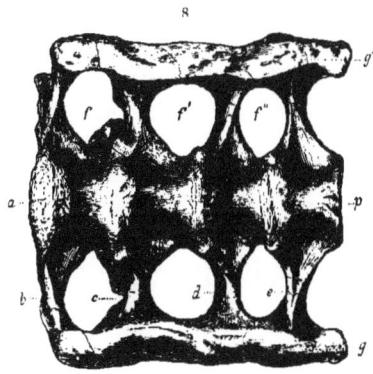

MOROSAURUS GRANDIS Marsh

Jurassic

PLATE XXXII.

PLATE XXXII.

Jurassic Dinosaurs.—Sauropoda.

Morosauridæ

    b, ball; *c*, cup; *d*, diapophysis; *f*, foramen in centrum; *m*, metapophysis; *n*, neural canal; *ns*, neural suture; *s*, neural spine; *x*, hyposphene; *z*, anterior zygapophysis; *z'*, posterior zygapophysis.

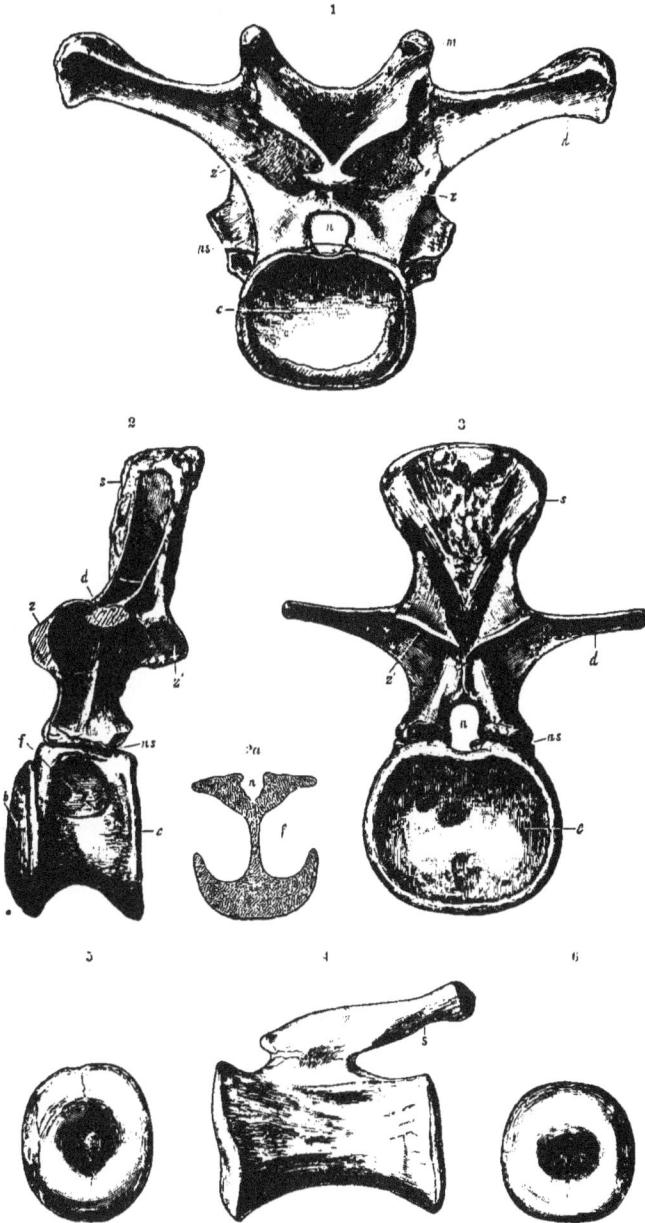

MOROSAURUS.

Jurassic.

PLATE XXXIII.

PLATE XXXIII.

JURASSIC DINOSAURS.—SAUROPODA.

MOROSAURIDÆ.

  a, anterior face of centrum of first sacral vertebra; *b*, transverse process of first vertebra; *c*, transverse process of second vertebra; *d*, transverse process of third vertebra; *e*, transverse process of last sacral vertebra; *f*, *f'*, *f''*, foramina between transverse processes of sacral vertebræ; *g*, surface for union with ilium; *n*, neural canal; *ns*, neural spine; *p*, posterior face of centrum of last sacral vertebra; *s*, suture; *z'*, posterior zygapophysis.

Both figures are one-fifth natural size.

310

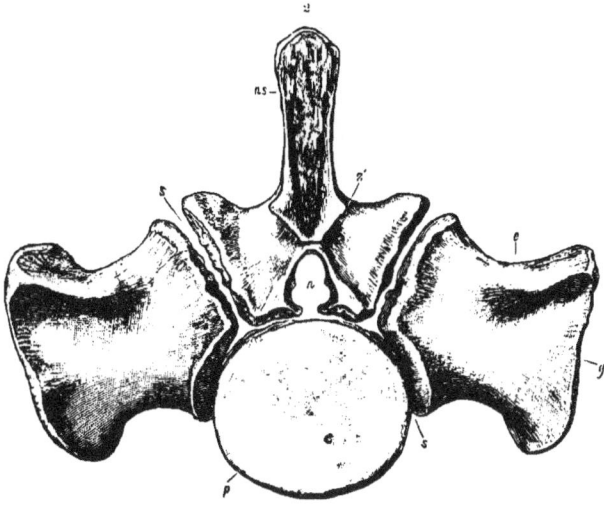

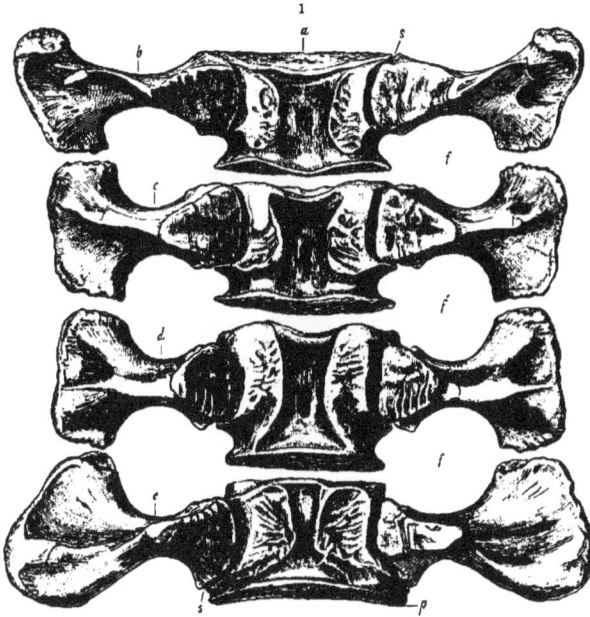

MOROSAURUS LENTUS Marsh,
Jurassic.

PLATE XXXIV.

PLATE XXXIV.

JURASSIC DINOSAURS.—SAUROPODA.

MOROSAURIDÆ.

312

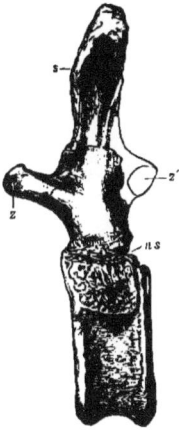

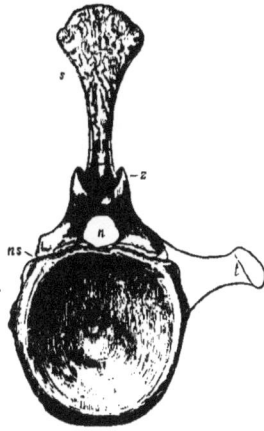

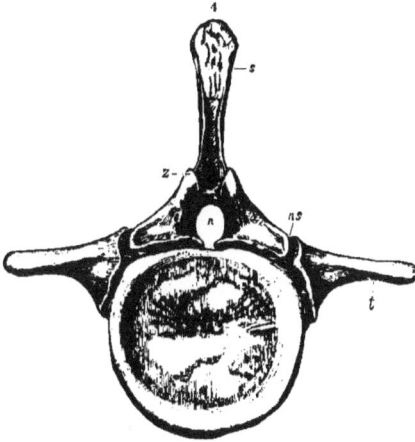

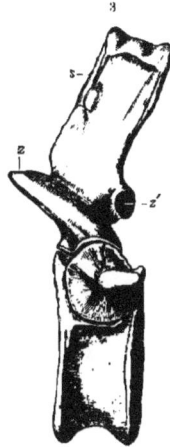

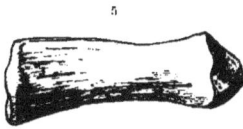

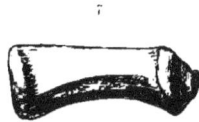

MOROSAURUS.
Jurassic.

PLATE XXXV.

PLATE XXXV.

Jurassic Dinosaurs.—Sauropoda.

Morosauridæ.

314

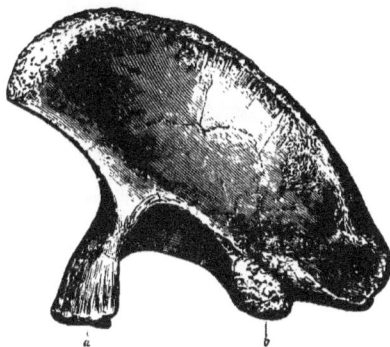

2

3

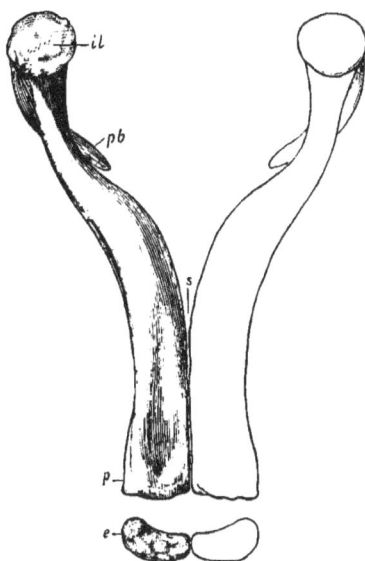

4

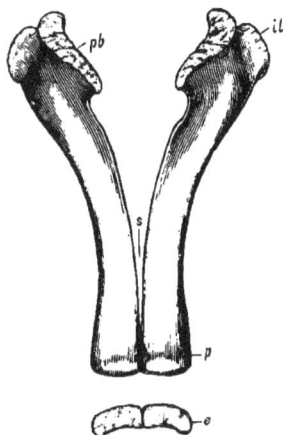

MOROSAURUS.
Jurassic.

PLATE XXXVI.

PLATE XXXVI.

JURASSIC DINOSAURS.—SAUROPODA.

MOROSAURIDÆ AND ATLANTOSAURIDÆ.

    a, acetabulum; *f,* foramen in pubis; *il,* ilium; *is,* ischium; *p,* pubis.

316

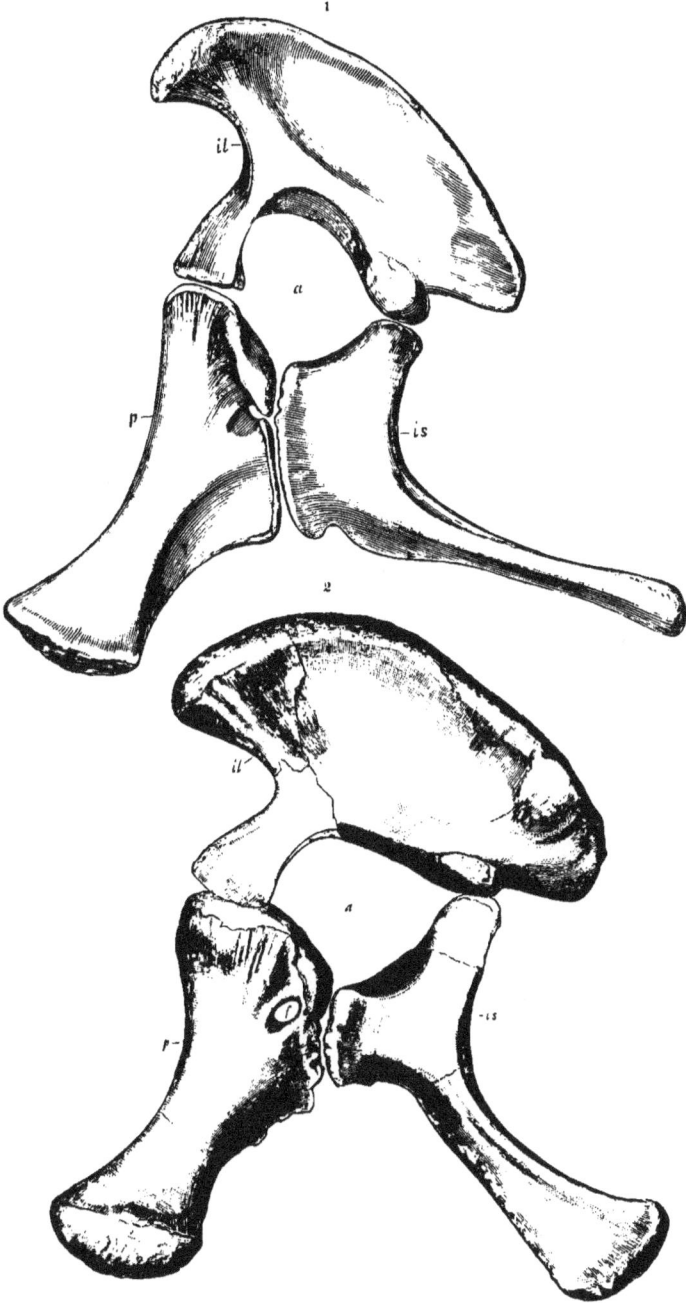

MOROSAURUS AND APATOSAURUS.
Jurassic.

PLATE XXXVII.

PLATE XXXVII.

JURASSIC DINOSAURS.—SAUROPODA.

MOROSAURIDÆ.

    318

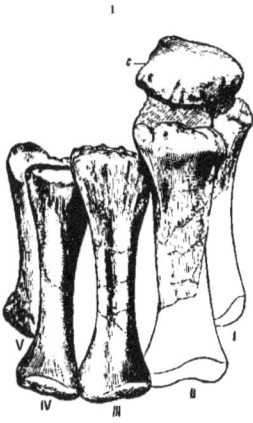

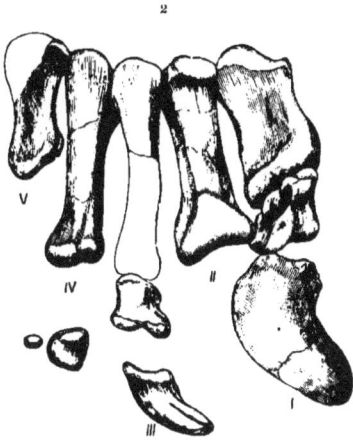

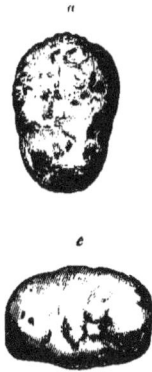

MOROSAURUS.
Jurassic.

PLATE XXXVIII.

PLATE XXXVIII.

JURASSIC DINOSAURS.—SAUROPODA.

MOROSAURIDÆ.

    Both figures are one-twentieth natural size.

   320

1

2

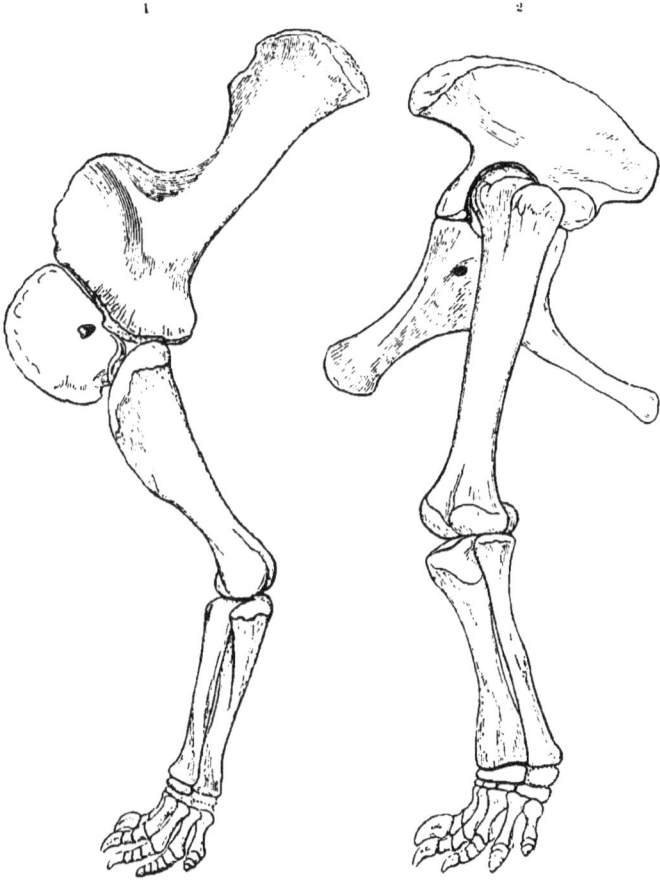

MOROSAURUS GRANDIS.
Jurassic.

PLATE XXXIX.

PLATE XXXIX.

Jurassic Dinosaurs.—Sauropoda.

Atlantosauridæ, Morosauridæ, and Diplodocidæ.

a, front view; a', anterior end; b, side view; c, back view; d, top view; h, hæmal orifice; p, posterior end; r, face for vertebra.

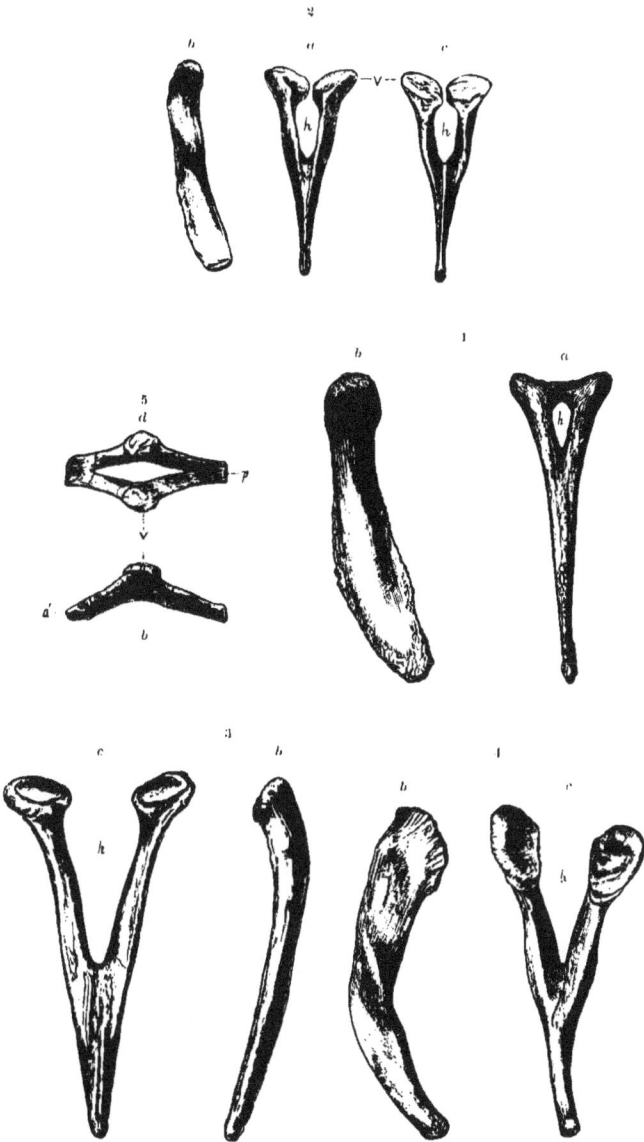

BRONTOSAURUS, APATOSAURUS, MOROSAURUS, AND DIPLODOCUS.
Jurassic.

PLATE XL.

323

PLATE XL.

JURASSIC DINOSAURS.—SAUROPODA.

PLEUROCŒLIDÆ.

    All the figures of vertebræ are one-half natural size.

    a, anterior end of centrum; *c*, face for chevron; *f*, foramen in centrum; *n*, neural canal; *p*, posterior end of centrum; *r*, face for rib; *s*, neural spine.

Potomac formation, Maryland.

324

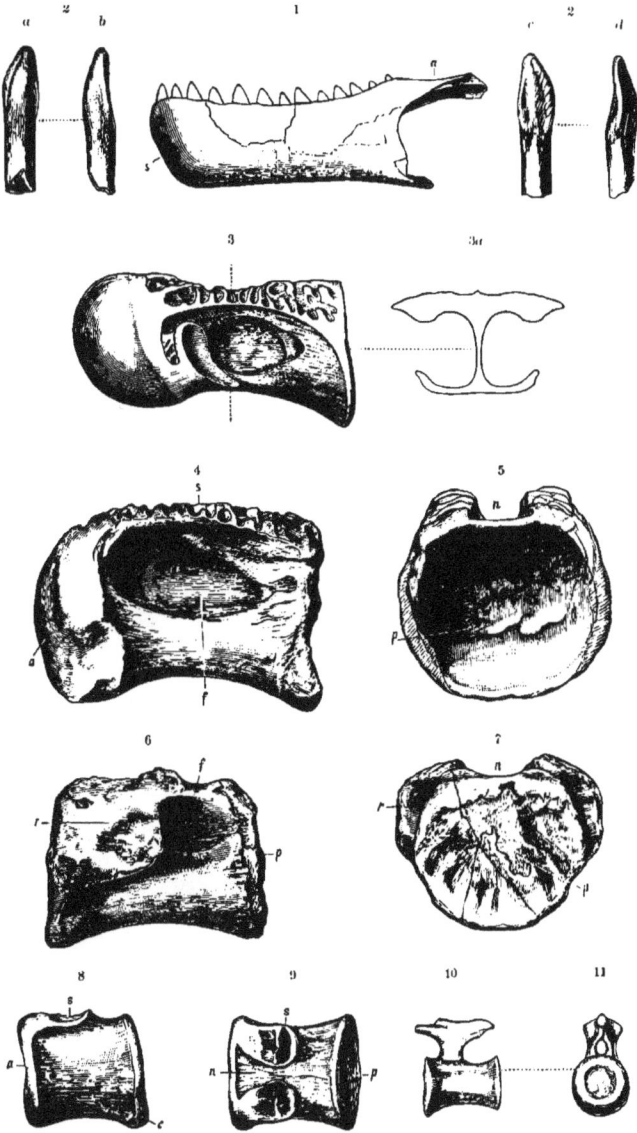

PLEUROCŒLUS NANUS Marsh.
Jurassic.

PLATE XLI.

325

PLATE XLI.

PLATE XLI.

Jurassic Dinosaurs.—Sauropoda.

Pleurocœlidæ.

All the figures are one-half natural size.

Potomac formation, Maryland.

326

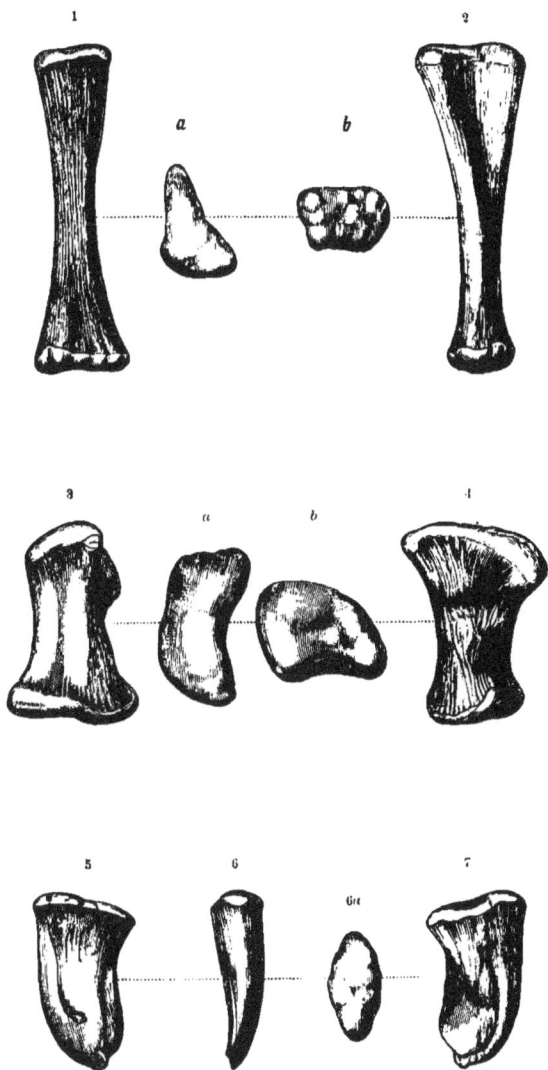

PLEUROCŒLUS NANUS.
Jurassic.

PLATE XLII.

PLATE XLII.

Jurassic Dinosaurs.—Sauropoda.

Atlantosauridæ.

328

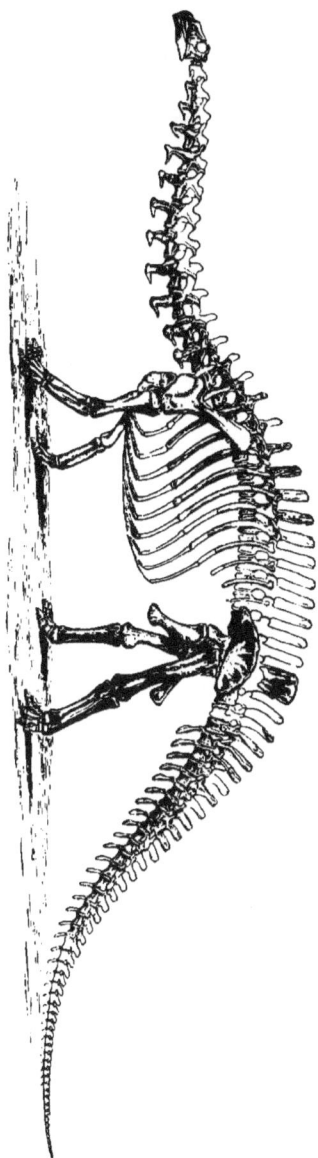

PLATE XLIII.

PLATE XLIII.

Jurassic Dinosaurs.—Predentata.

Stegosauridæ.

a, anterior nares; an, angular; ar, articular; b, orbit; c, lower temporal fossa; d, dentary; e, supratemporal fossa; f, frontal; fp, postfrontal; j, jugal; l, lachrymal; m, maxillary; n, nasal; oc, occipital condyle; os, supraoccipital; p, parietal; pd, predentary; pf, prefrontal; pm, premaxillary; po, postorbital; q, quadrate; s, splenial; sa, surangular; so, supraorbital; sq, squamosal.

All the figures are one-fourth natural size.

330

1

2

3

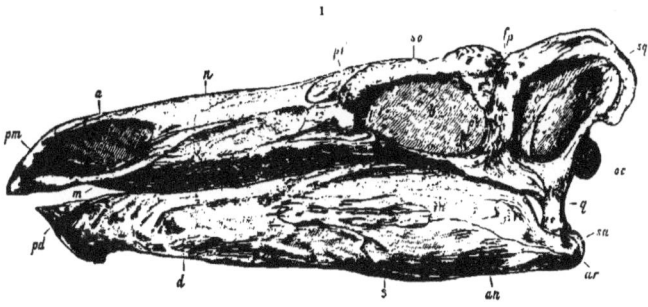

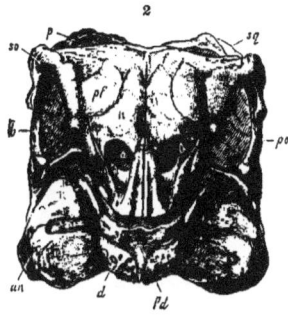

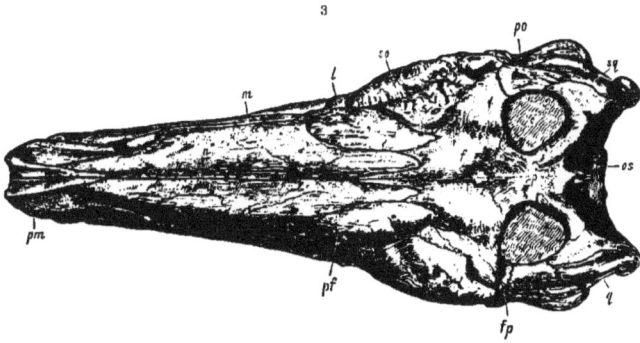

STEGOSAURUS STENOPS Marsh.
Jurassic.

PLATE XLIV.

PLATE XLIV.

Jurassic Dinosaurs.—Predentata.

Stegosauridæ.

332

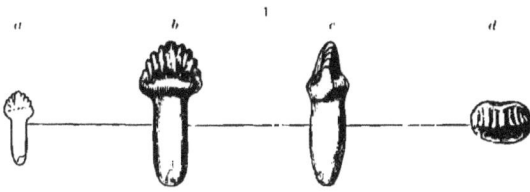

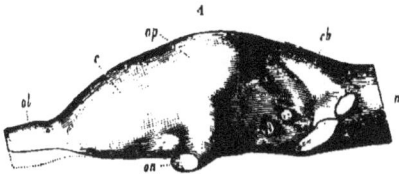

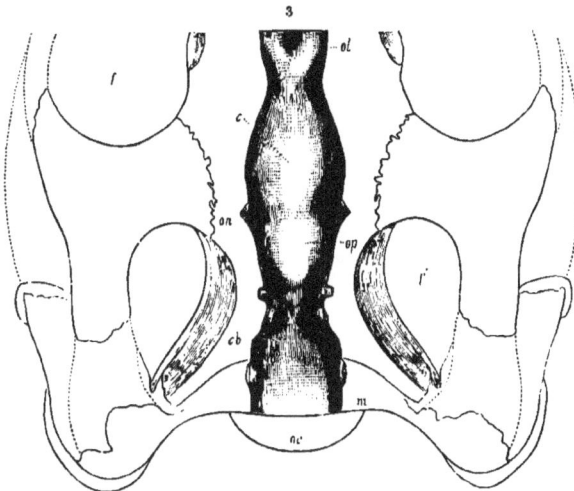

STEGOSAURUS AND PRICONODON.

Jurassic.

PLATE XLV.

PLATE XLV.

Jurassic Dinosaurs.—Predentata.

Stegosauridæ.

c, face for chevron; d, diapophysis; n, neural canal; p, parapophysis; s, neural spine; z, anterior zygapophysis; z', posterior zygapophysis.

All the figures are one-eighth natural size.

334

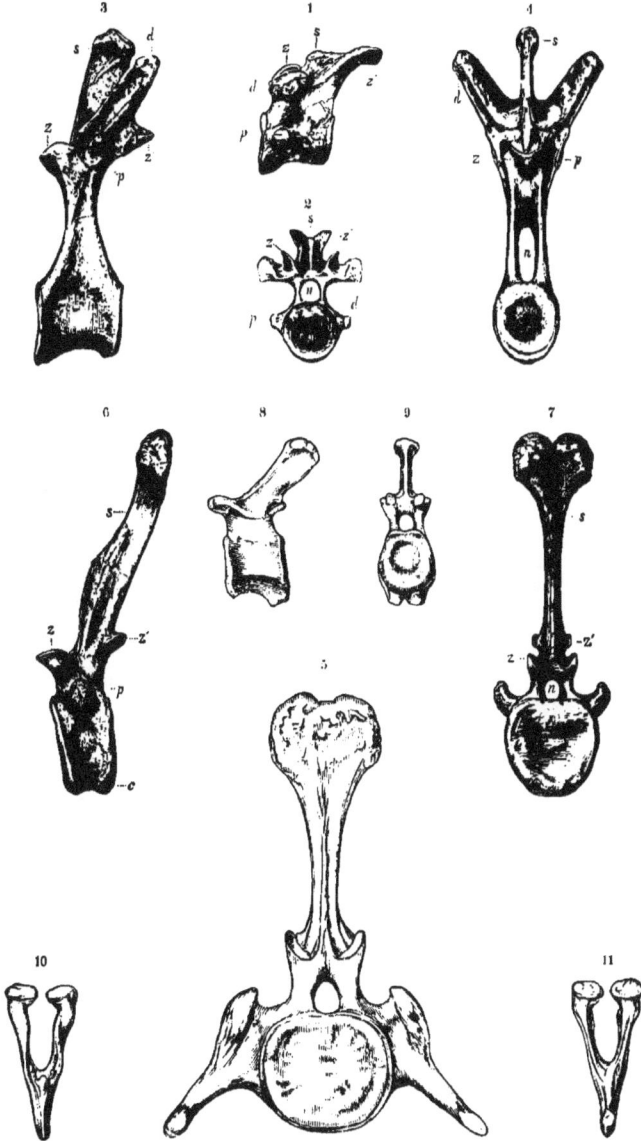

STEGOSAURUS UNGULATUS Marsh.

Jurassic.

PLATE XLVI.

PLATE XLVI.

Jurassic Dinosaurs.—Predentata.

Stegosauridæ.

336

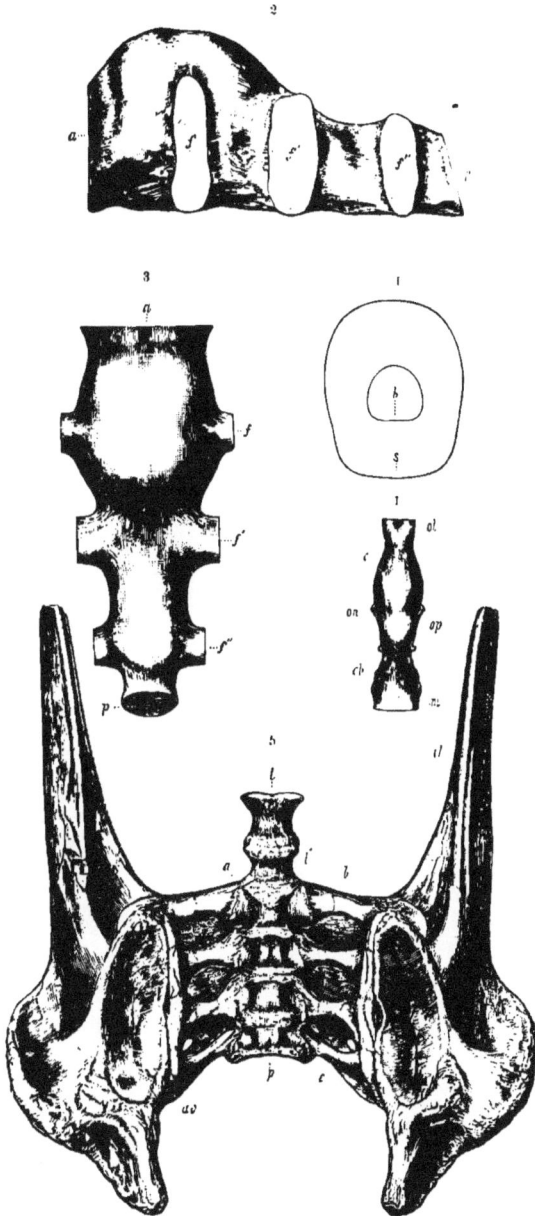

STEGOSAURUS UNGULATUS.

Jurassic.

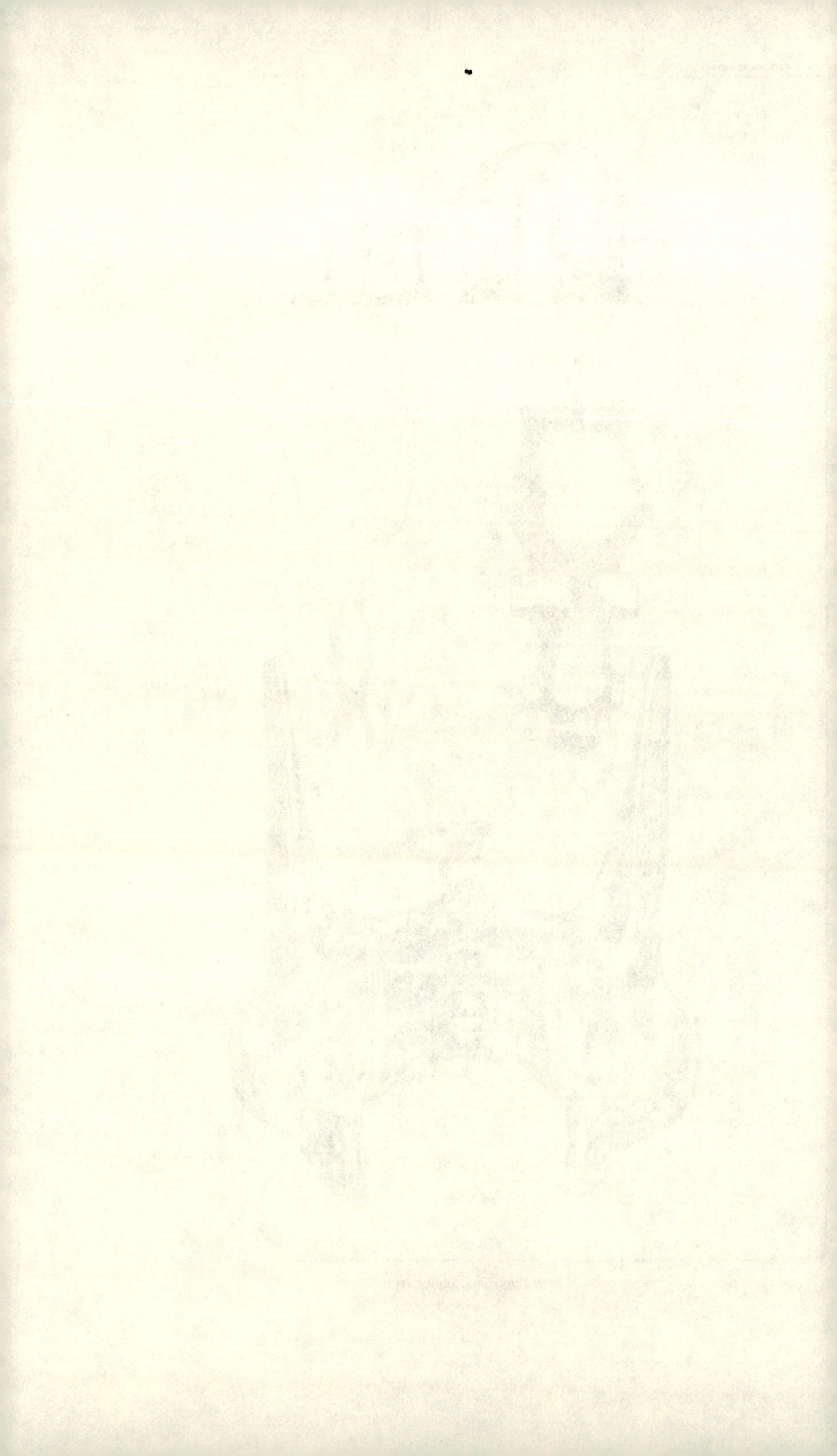

PLATE XLVII.

PLATE XLVII.

Jurassic Dinosaurs.—Predentata.

Stegosauridæ.

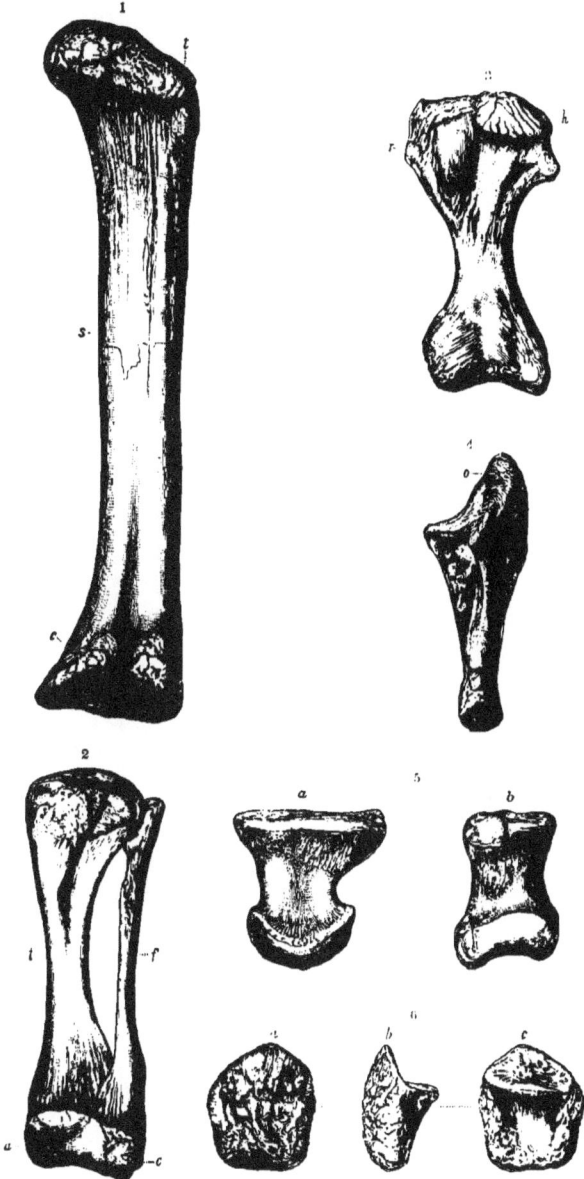

STEGOSAURUS UNGULATUS.
Jurassic

PLATE XLVIII.

PLATE XLVIII.

Jurassic Dinosaurs.—Predentata.

Stegosauridæ.

    a, acetabulum; *c*, coracoid; *f*, femur; *f*, fibula; *h*, humerus; *il*, ilium; *is*,
ischium; *p*, pubis; *p'*, postpubis; *r*, radius; *s*, scapula; *t*, tibia; *u*, ulna:
I, first digit; *V*, fifth digit.

340

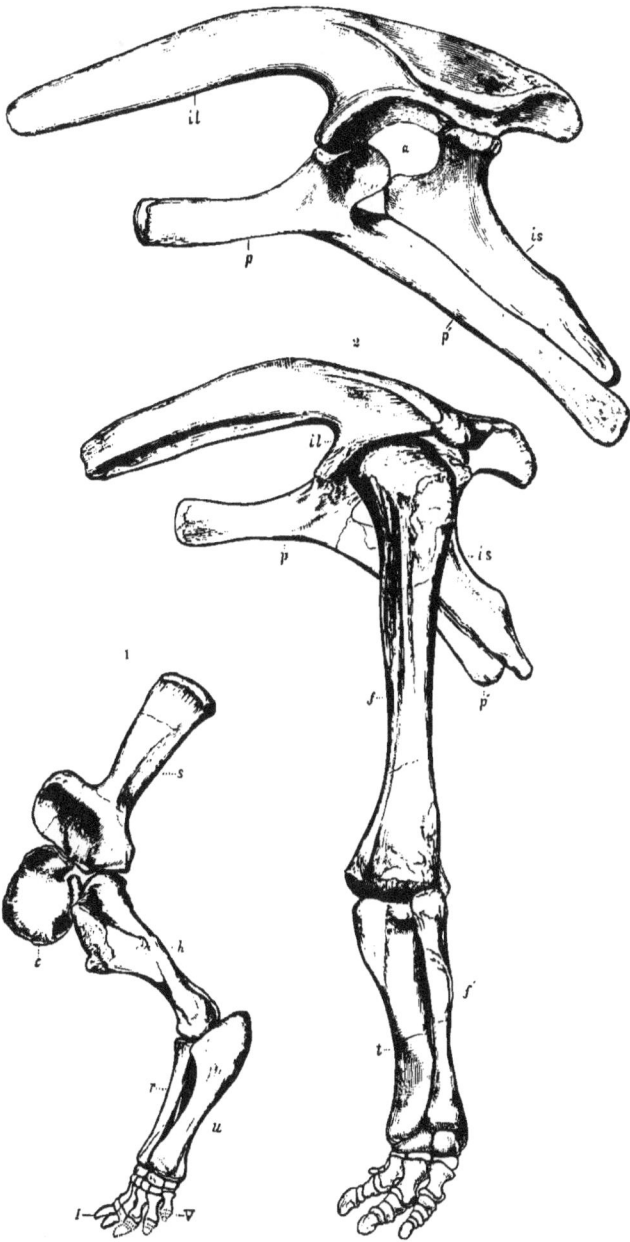

STEGOSAURUS

Jurassic.

PLATE XLIX.

PLATE XLIX.

JURASSIC DINOSAURS.—PREDENTATA.

STEGOSAURIDÆ.

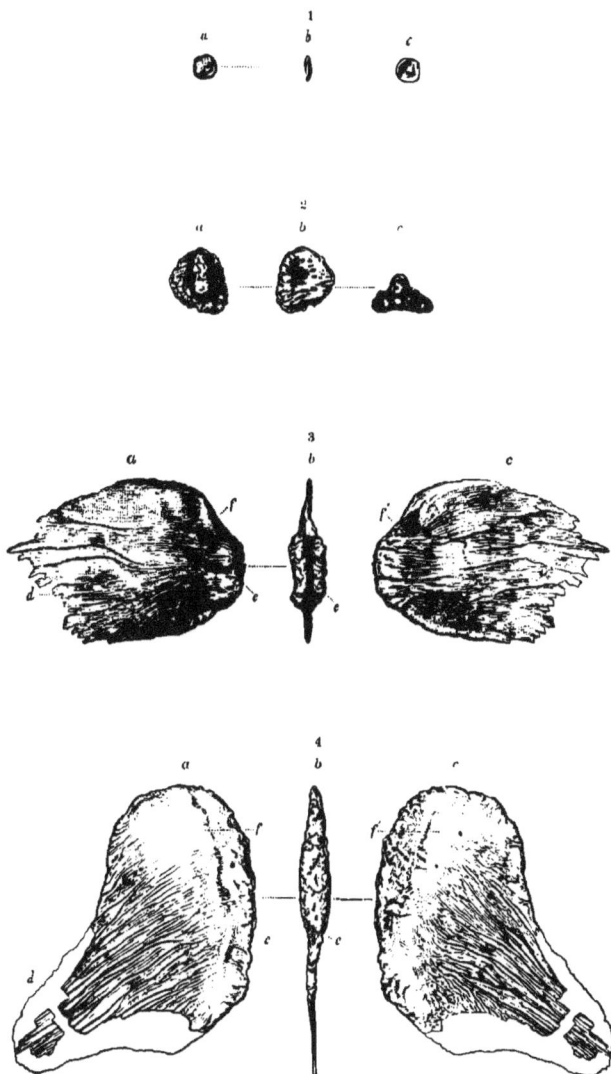

STEGOSAURUS UNGULATUS
Jurassic.

PLATE L.

PLATE L.

JURASSIC DINOSAURS.—PREDENTATA.

STEGOSAURIDÆ.

344

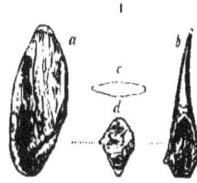

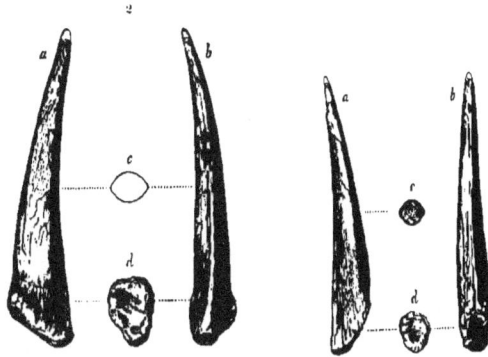

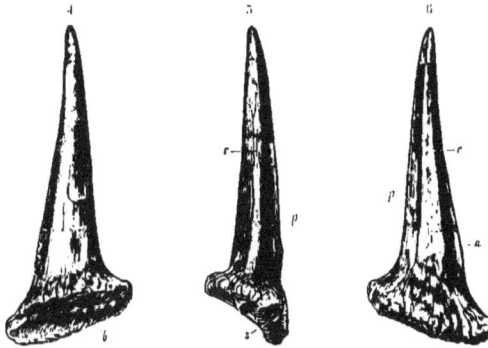

STEGOSAURUS.
Jurassic.

PLATE LI.

PLATE LI.

Jurassic Dinosaurs.—Predentata.

Stegosauridæ.

The bones represented are essentially in the position in which they were found.

346

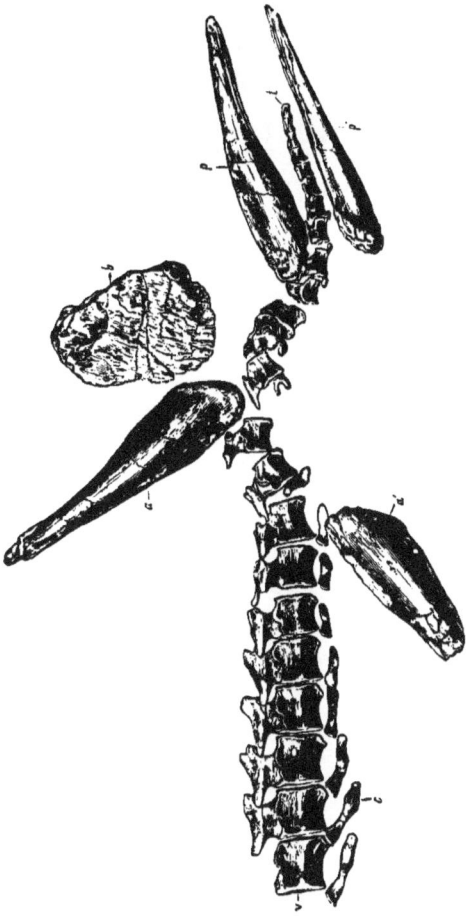

PLATE LII.

PLATE LII.

Jurassic Dinosaurs.—Predentata.

Stegosauridæ.

348

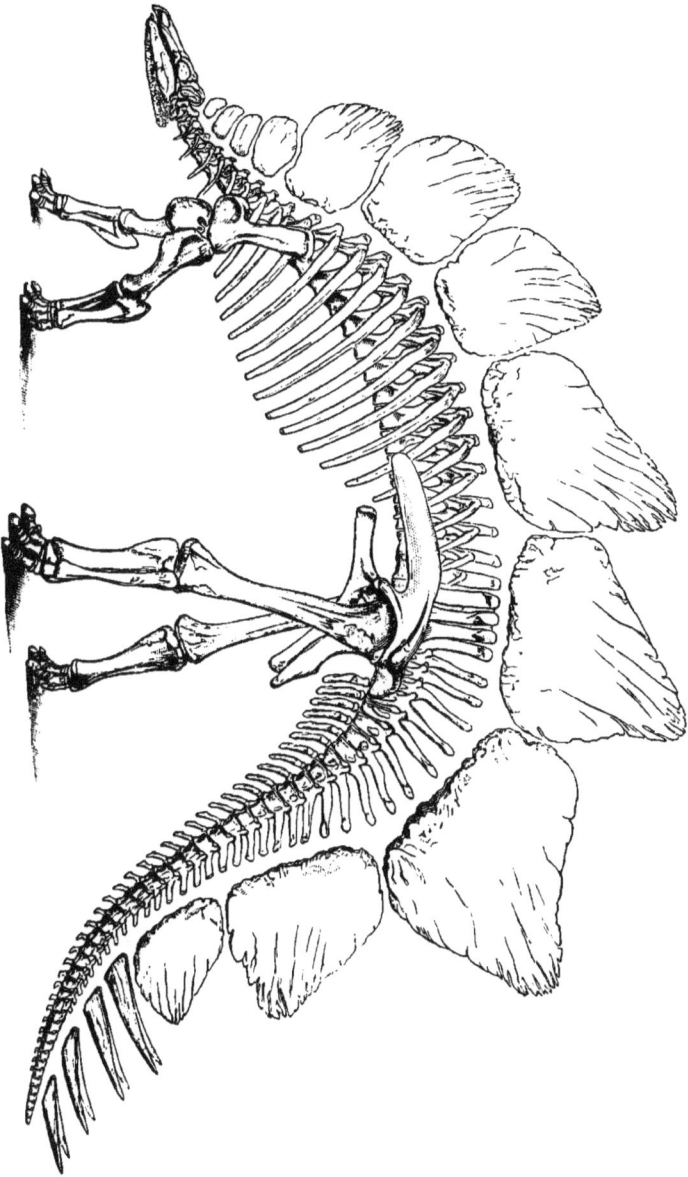

PLATE LIII.

PLATE LIII.

JURASSIC DINOSAURS.—PREDENTATA.

CAMPTOSAURIDÆ.

350

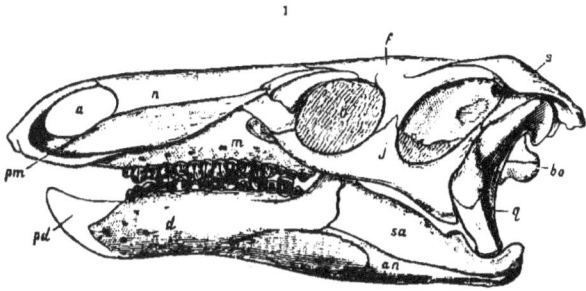

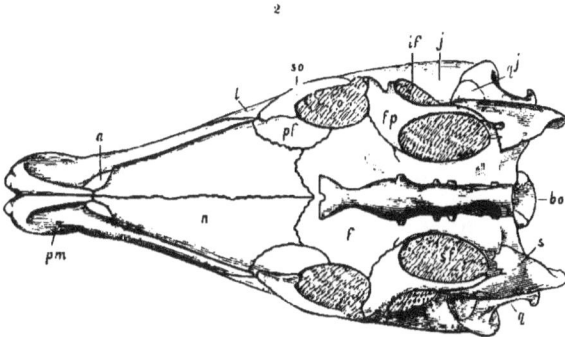

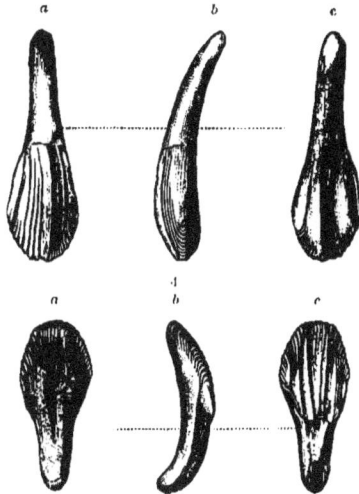

CAMPTOSAURUS MEDIUS Marsh.

Jurassic.

PLATE LIV.

PLATE LIV.

JURASSIC DINOSAURS.—PREDENTATA.

CAMPTOSAURIDÆ.

352

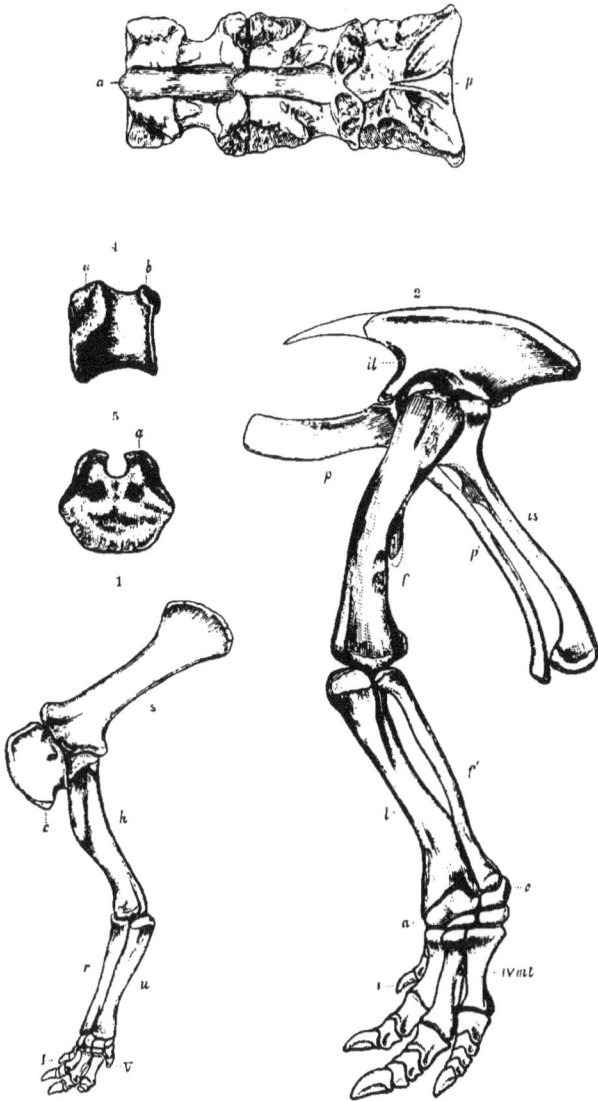

CAMPTOSAURUS DISPAR Marsh.
Jurassic.

PLATE LV.

PLATE LV.

JURASSIC DINOSAURS.—PREDENTATA.

CAMPTOSAURIDÆ AND LAOSAURIDÆ.

354

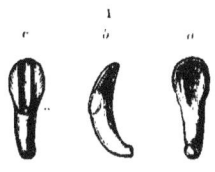

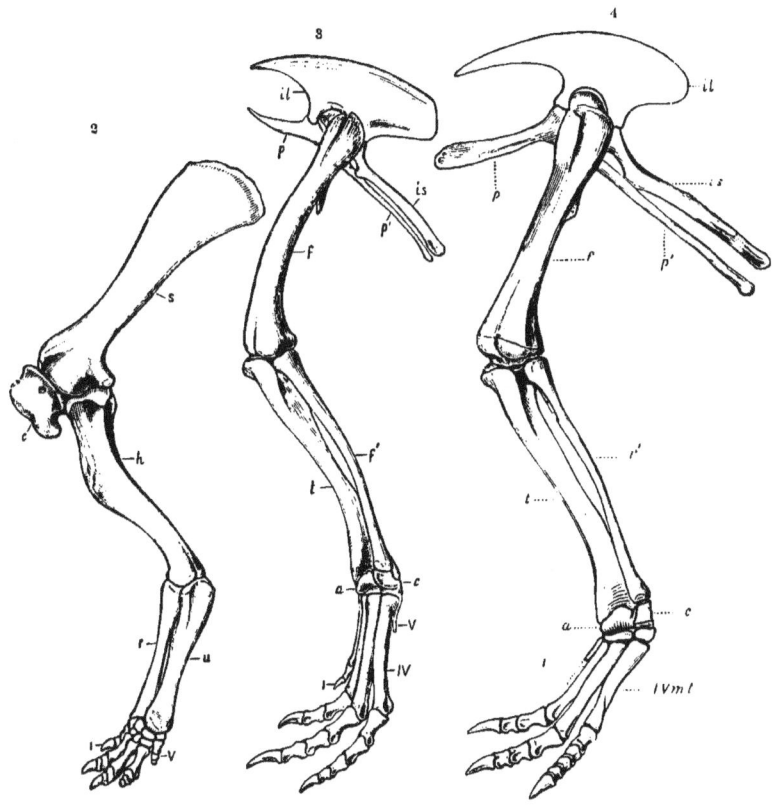

CAMPTOSAURUS, DRYOSAURUS, AND LAOSAURUS.

Jurassic.

PLATE LXIV.

CRETACEOUS DINOSAURS.—PREDENTATA.

CERATOPSIDÆ.

a, anterior face of centrum; *c*, face for chevron; *h*, facet for head of rib; *n*, neural canal; *p*, posterior face of centrum; *r*, rib; *s*, neural spine; *t*, facet for tubercle of rib; *z*, anterior zygapophysis; *z'*, posterior zygapophysis.

All the figures are one-eighth natural size.

PLATE LXIV.

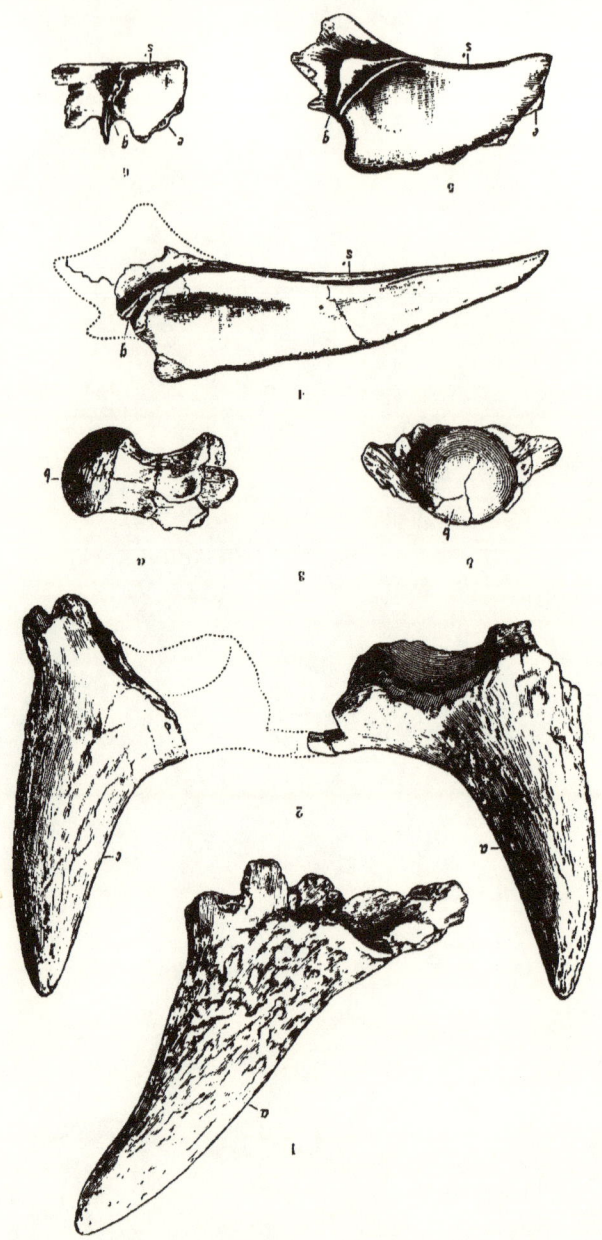

PLATE LXIII.

Cretaceous Dinosaurs.—Predentata.

Ceratopsidæ.

370

PLATE LXIII.

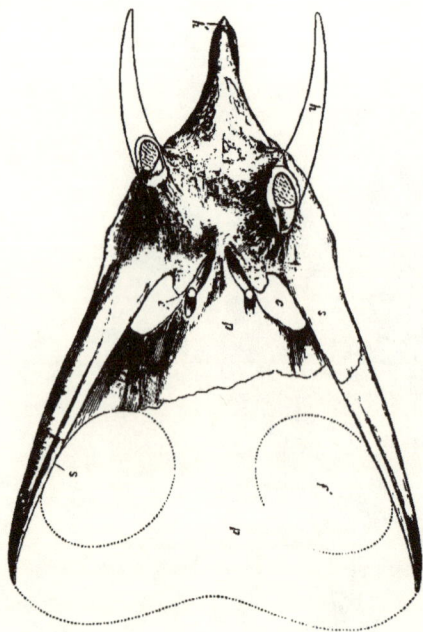

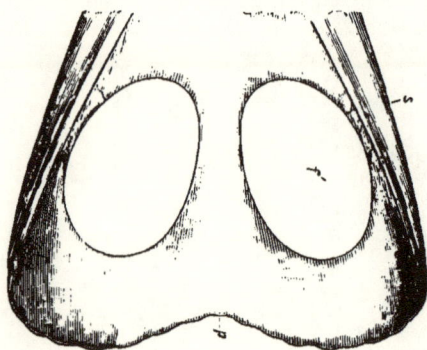

TOROSAURUS
Cretaceous

PLATE LXII.

CRETACEOUS DINOSAURS.—PREDENTATA.

CERATOPSIDÆ.

PLATE LXII.

PLATE LVI.

PLATE LVI

Jurassic Dinosaurs.—Predentata.

Camptosauridæ.

356

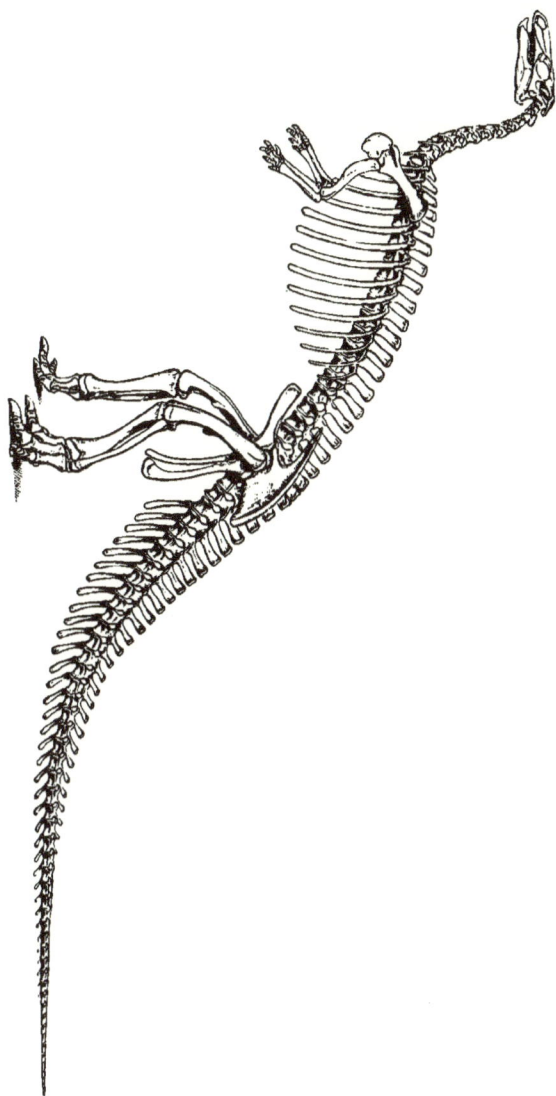

PLATE LVII.

PLATE LVII.

JURASSIC DINOSAURS.—PREDENTATA.

LAOSAURIDÆ.

358

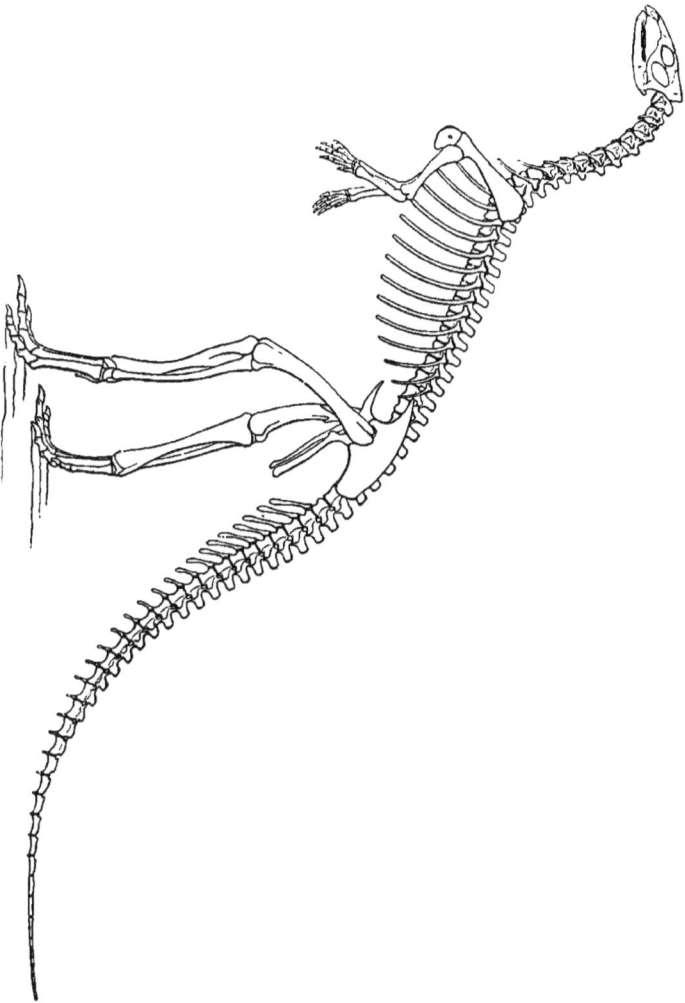

PLATE LVIII.

PLATE LVIII.

Cretaceous Dinosaurs.—Theropoda.

Ornithomimidæ.

360

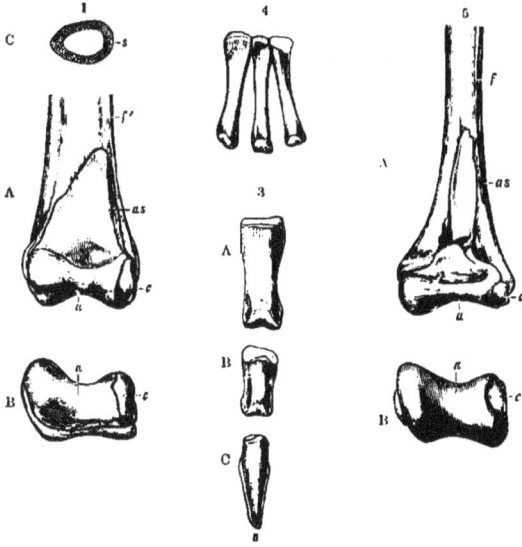

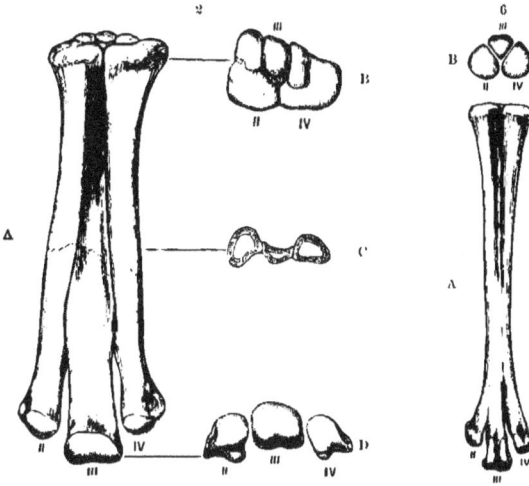

ORNITHOMIMUS VELOX Marsh
Cretaceous

PLATE LIX.

PLATE LIX.

Cretaceous Dinosaurs.—Predentata.

Ceratopsidæ.

362

TRICERATOPS PRORSUS Marsh

Cretaceous

PLATE LX.

PLATE LX.

Cretaceous Dinosaurs.—Predentata.

CERATOPSIDÆ.

All the figures are one-twentieth natural size.

364

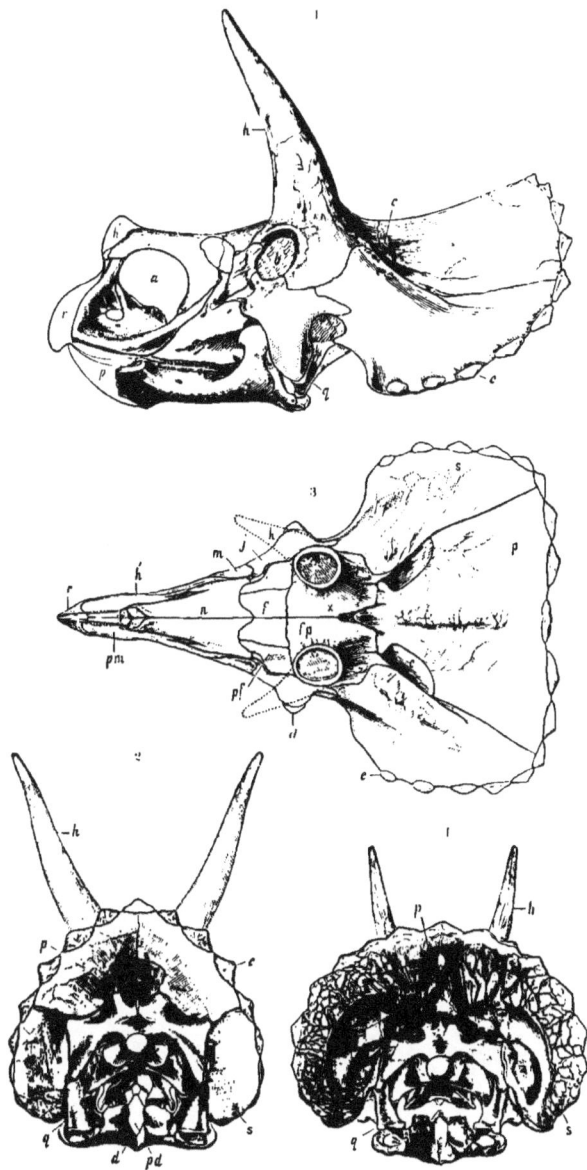

STERRHOLOPHUS AND TRICERATOPS.
Cretaceous

PLATE LXI.

PLATE LXI.

CRETACEOUS DINOSAURS.—PREDENTATA.

CERATOPSIDÆ.

366

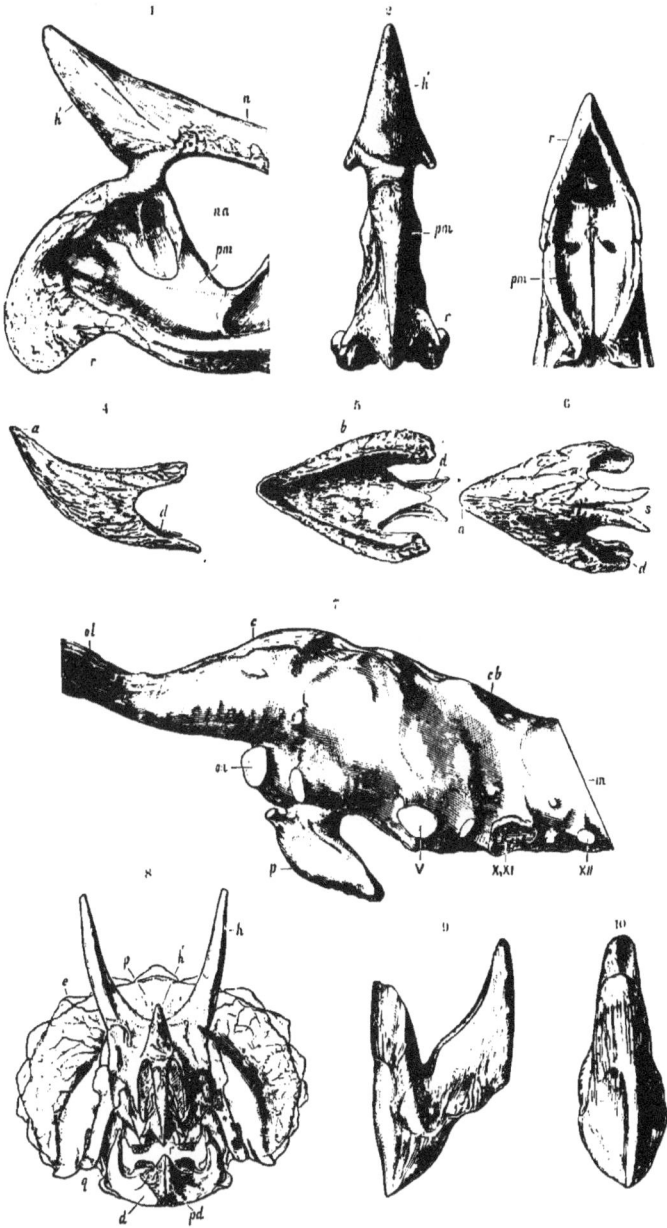

TRICERATOPS.
Cretaceous

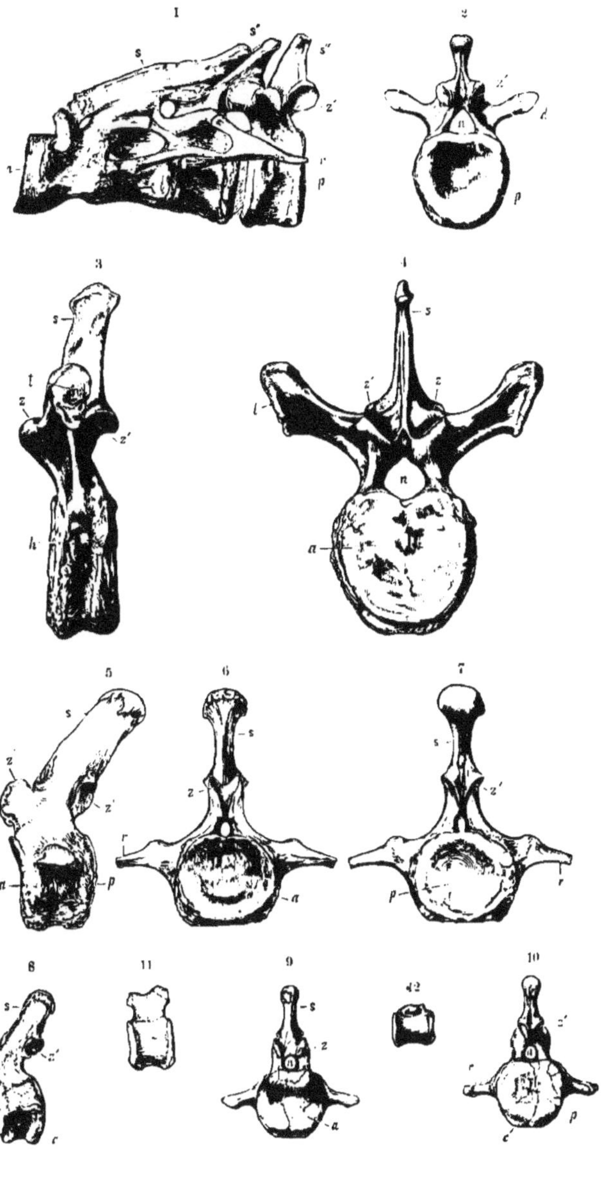

PLATE LXV.

PLATE LXV.

CERATOPSIDÆ.

Page.

Sacrum of *Triceratops prorsus* Marsh; seen from below; one-eighth natural size.   212
    a, anterior face of first sacral vertebra; *p*, posterior face of last sacral vertebra; *s*, neural spine of last vertebra; *z*, anterior zygapophysis of first vertebra; *1-10*, transverse processes, left side.

374

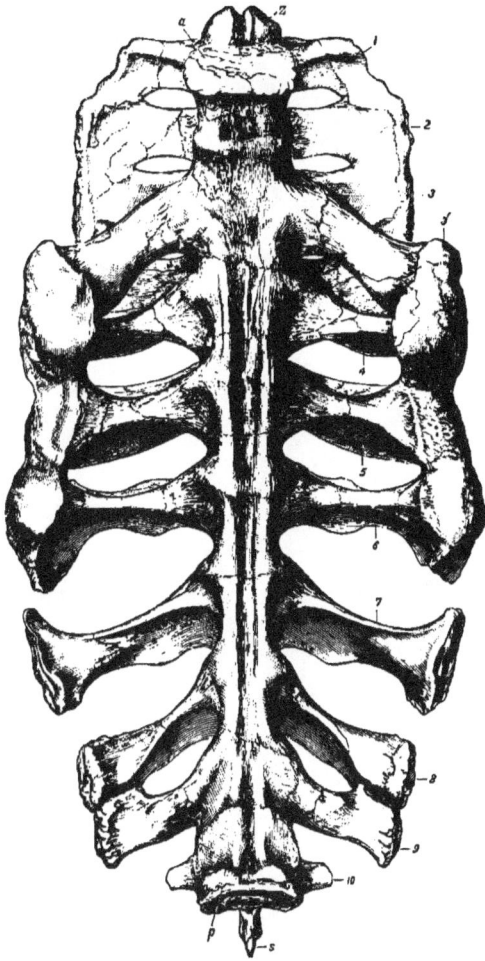

TRICERATOPS PRORSUS.
Cretaceous.

PLATE LXVI.

PLATE LXVI.

CRETACEOUS DINOSAURS.—PREDENTATA.

CERATOPSIDÆ.

    cr, coracoid; *g*, glenoid fossa; *h*, head; *o*, olecranon process; *r*, radial crest; *r'*, face for radius; *s*, suture; *sc*, scapula.

    All the figures are one-eighth natural size.

376

1

h

r

3

o

sc

g

s

cr

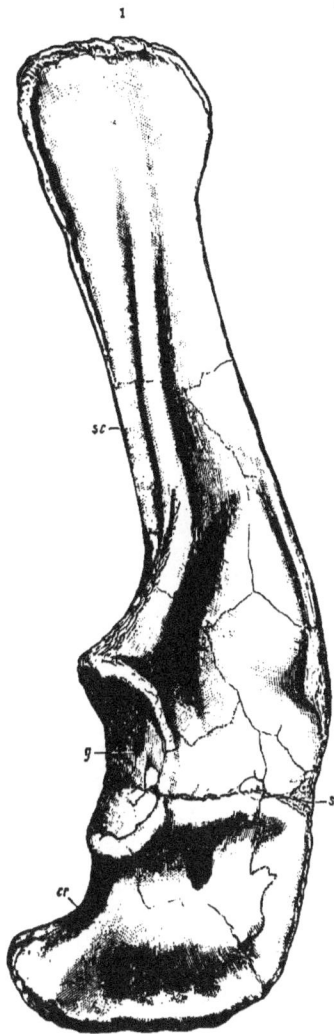

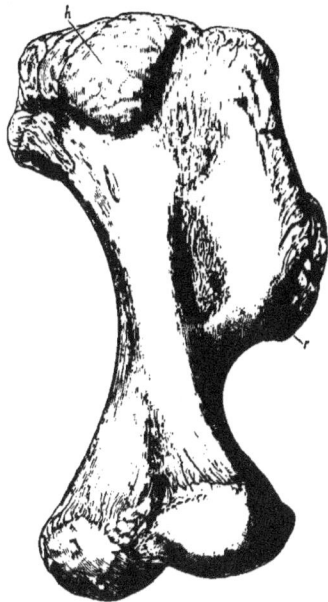

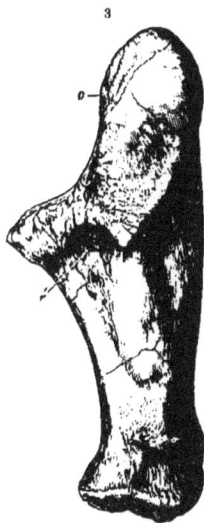

TRICERATOPS PRORSUS
Cretaceous.

PLATE LXVII.

PLATE LXVII.

CRETACEOUS DINOSAURS.—PREDENTATA.

CERATOPSIDÆ.

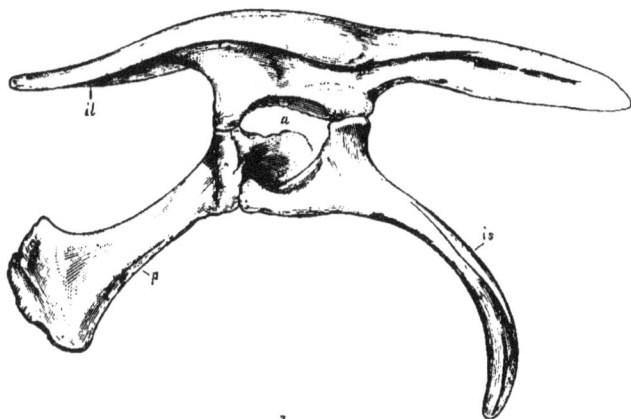

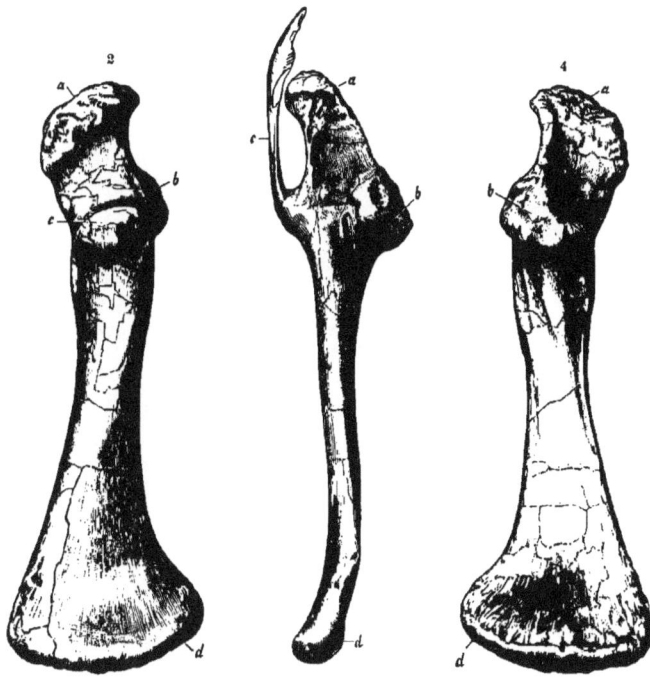

STERRHOLOPHUS AND TRICERATOPS.
Cretaceous

PLATE LXVIII.

PLATE LXVIII.

CRETACEOUS DINOSAURS.—PREDENTATA.

CERATOPSIDÆ.

a, astragalus; c, inner condyle; c', cnemial crest; f, face for fibula; h, head; t, great trochanter.

All the figures are one-eighth natural size.

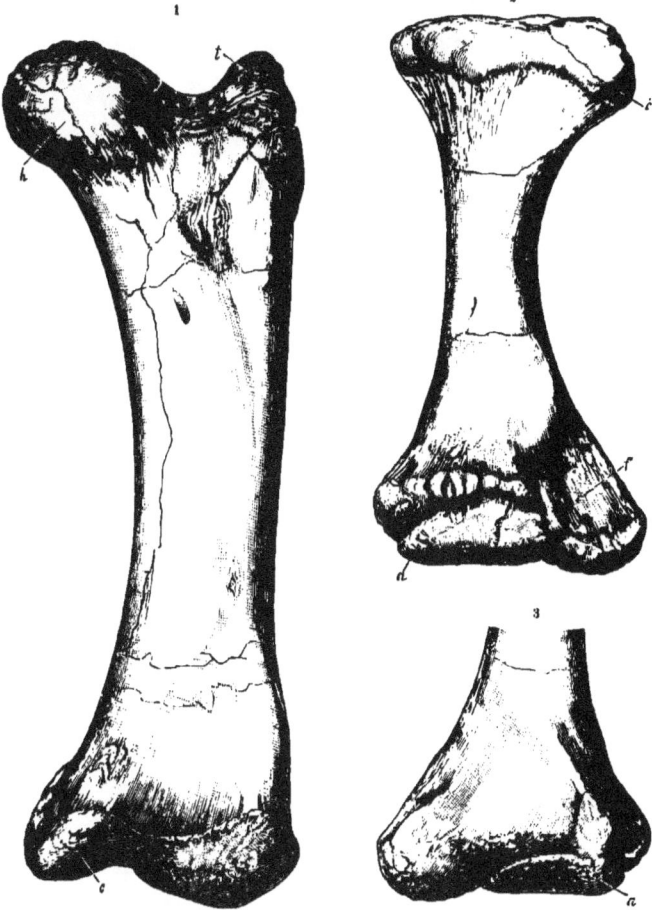

TRICERATOPS PRORSUS.
Cretaceous.

PLATE LXIX.

PLATE LXIX.

CRETACEOUS DINOSAURS.—PREDENTATA.

CERATOPSIDÆ.

382

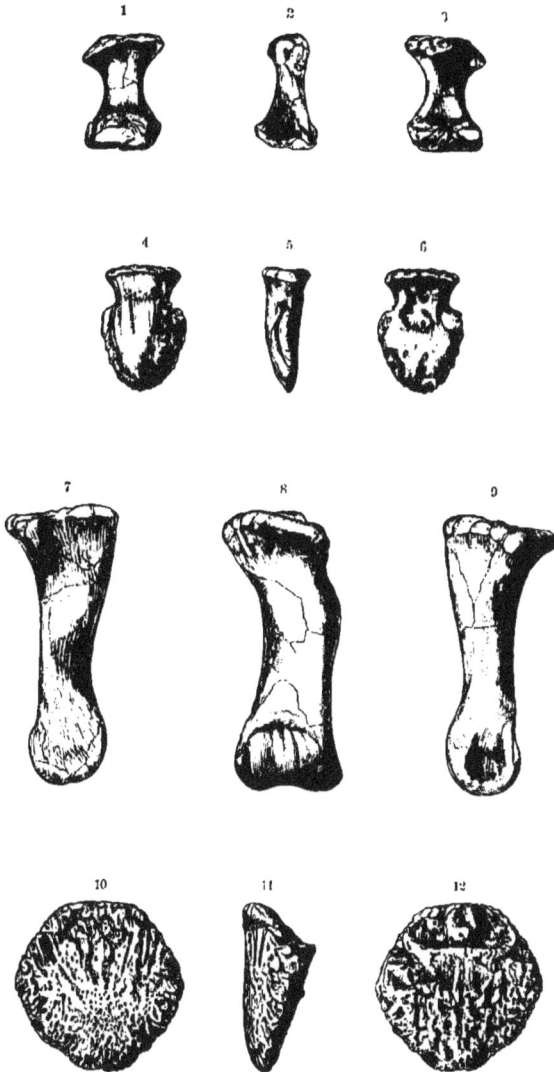

STERRHOLOPHUS AND TRICERATOPS.

Cretaceous.

PLATE LXX.

PLATE LXX.

CRETACEOUS DINOSAURS.—PREDENTATA.

CERATOPSIDÆ.

384

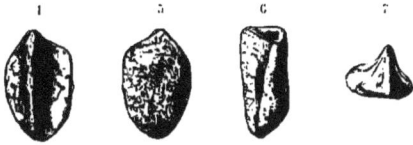

TRICERATOPS.
Cretaceous

PLATE LXXI.

PLATE LXXI.

CRETACEOUS DINOSAURS.—PREDENTATA.

CERATOPSIDÆ.

386

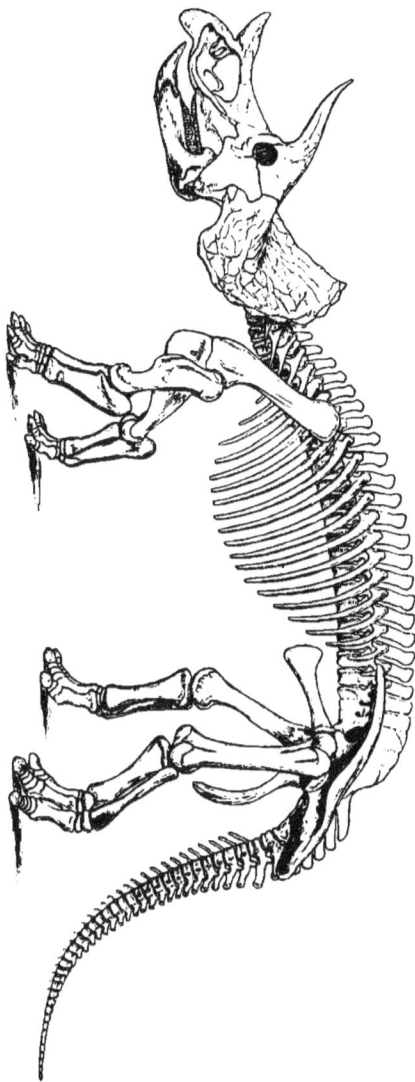

PLATE LXXII.

PLATE LXXII.

CRETACEOUS DINOSAURS.—PREDENTATA.

CLAOSAURIDÆ.

All the figures are one-tenth natural size.

388

1

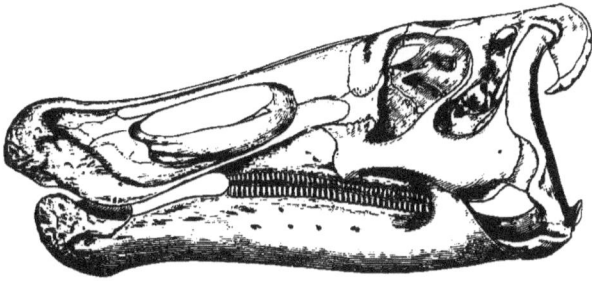

2

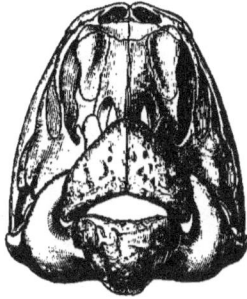

3

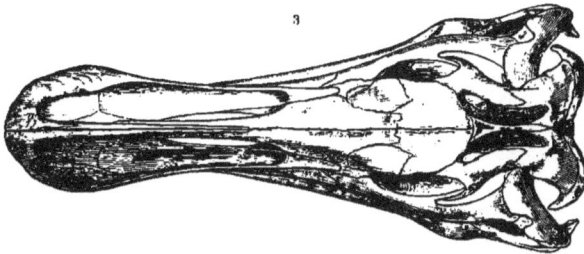

CLAOSAURUS ANNECTENS Marsh.
Cretaceous.

PLATE LXXIII.

PLATE LXXIII.

CRETACEOUS DINOSAURS.—PREDENTATA.

CLAOSAURIDÆ.

390

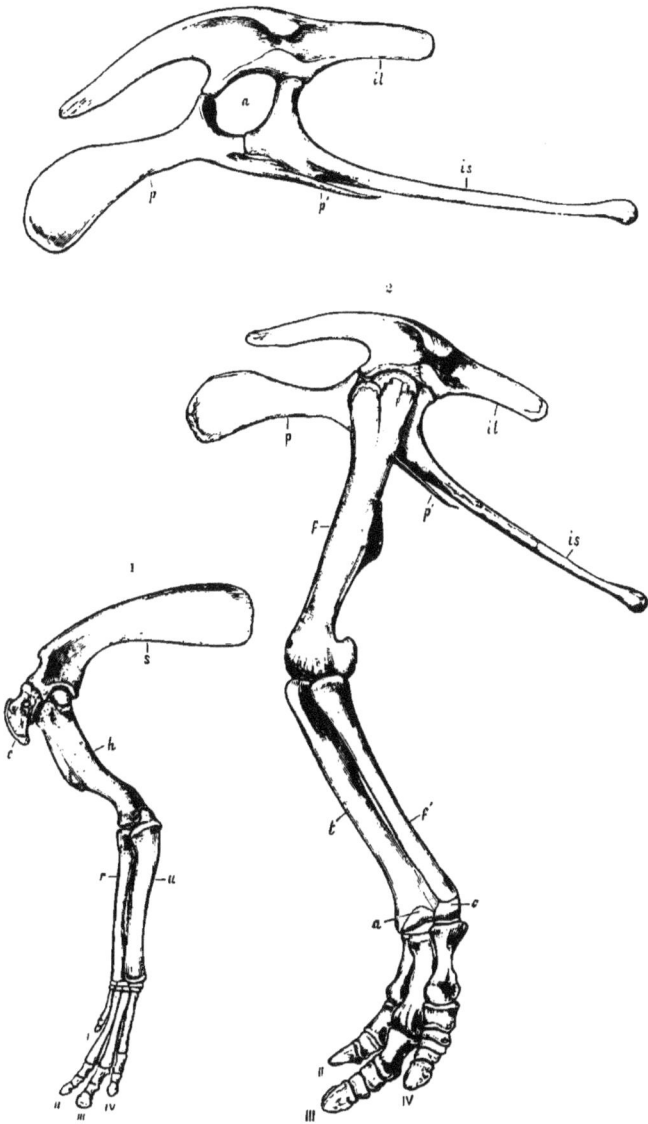

CLAOSAURUS ANNECTENS.
Cretaceous.

PLATE LXXIV.

PLATE LXXIV.

CRETACEOUS DINOSAURS.—PREDENTATA.

CLAOSAURIDÆ.

392

PLATE LXXV.

PLATE LXXV.

CRETACEOUS DINOSAURS.—PREDENTATA.

394

PLATE LXXVI.

PLATE LXXVI.

Dinosaurian Skulls; showing size of brain.

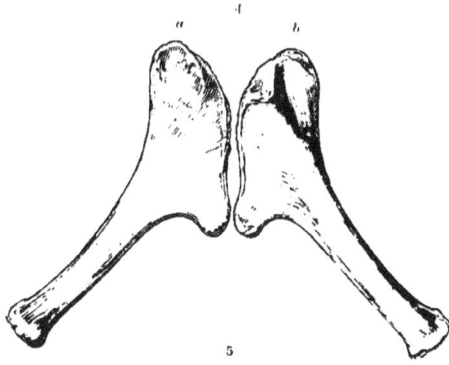

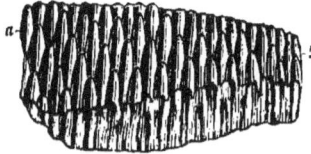

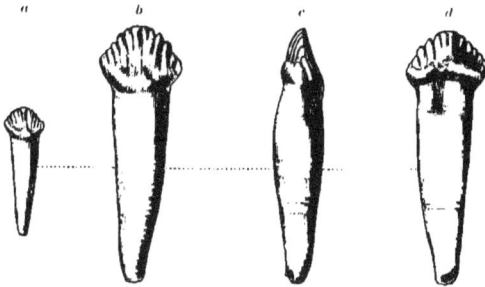

CLAOSAURUS, TRACHODON, NODOSAURUS, AND PALÆOSCINCUS.
Cretaceous

PLATE LXXVII.

PLATE LXXVII.

Brain casts of dinosaurs.

c, cerebral hemispheres; cb, cerebellum; m, medulla; ol, olfactory lobe; on, optic nerve; op, optic lobe; p, pituitary body; V, fifth nerve; X, XI, tenth and eleventh nerves; XII, twelfth nerve.

398

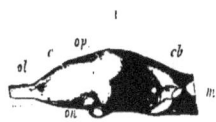

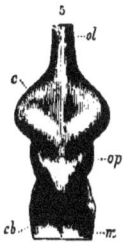

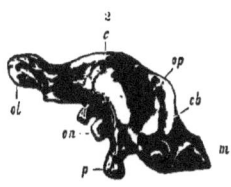

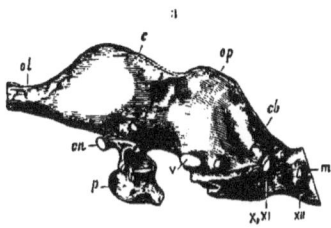

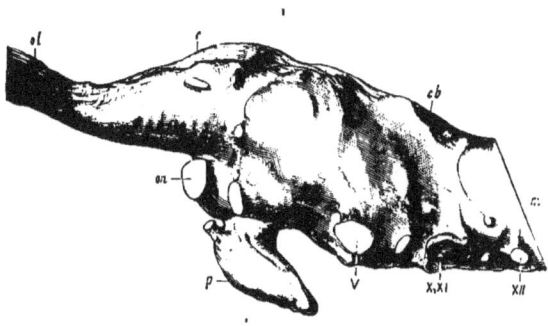

BRAIN CASTS OF DINOSAURS. CERATOSAURUS, CLAOSAURUS, STEGOSAURUS, TRICERATOPS, AND RECENT ALLIGATOR

PLATE LXXVIII.

PLATE LXXVIII.

Teeth of Predentate Dinosaurs.

400

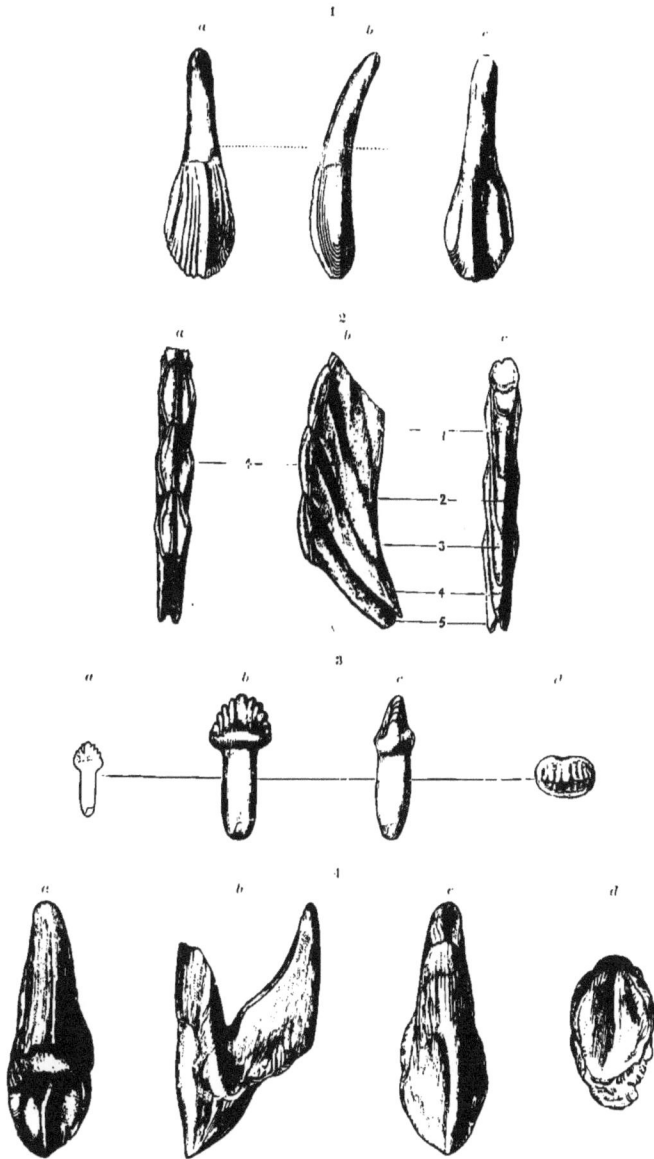

TEETH OF PREDENTATE DINOSAURS CAMPTOSAURUS, CLAOSAURUS, STEGOSAURUS, AND TRICERATOPS

PLATE LXXIX.

PLATE LXXIX.

      p, prepubis; *p'*, postpubis.

402

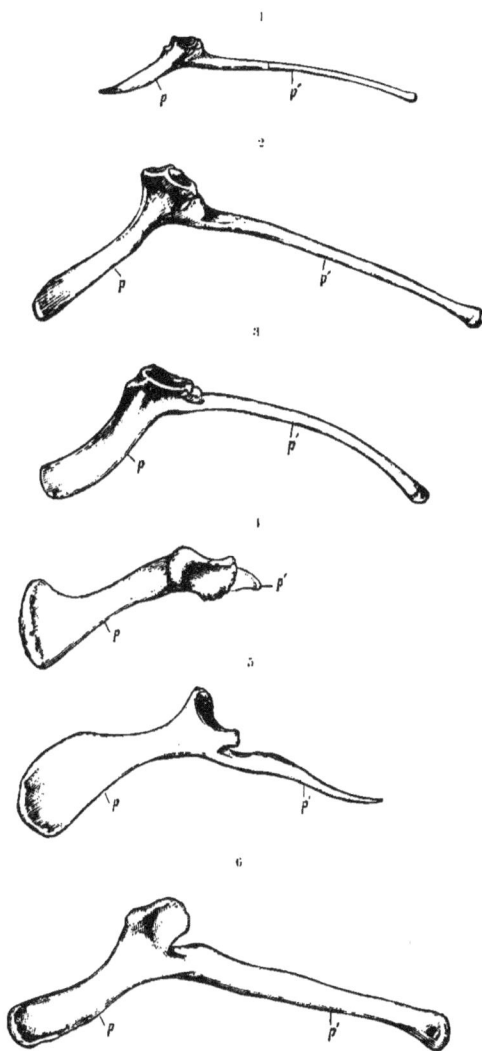

PUBES OF PREDENTATE DINOSAURS. CAMPTOSAURUS, CLAOSAURUS, DRYOSAURUS, LAOSAURUS STEGOSAURUS, AND TRICERATOPS.

PLATE LXXX.

PLATE LXXX.

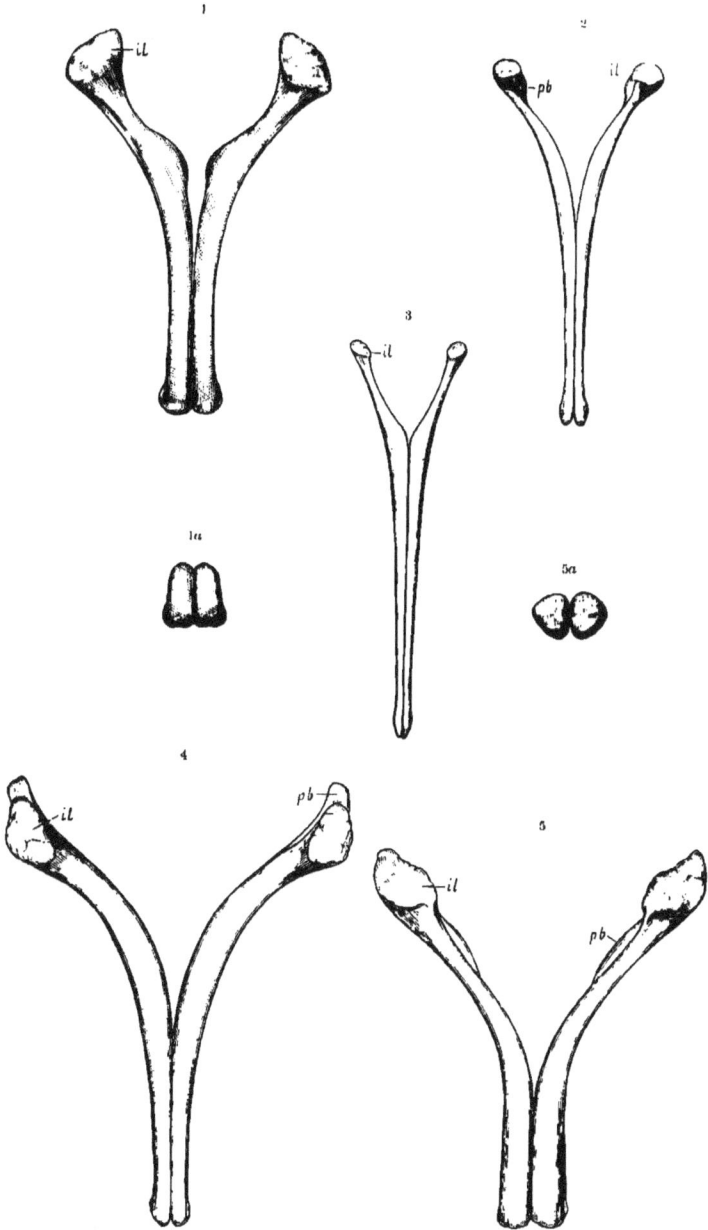

ISCHIA OF PREDENTATE DINOSAURS. CAMPTOSAURUS, CLAOSAURUS, DRYOSAURUS, STEGOSAURUS, AND TRICERATOPS

PLATE LXXXI.

PLATE LXXXI.

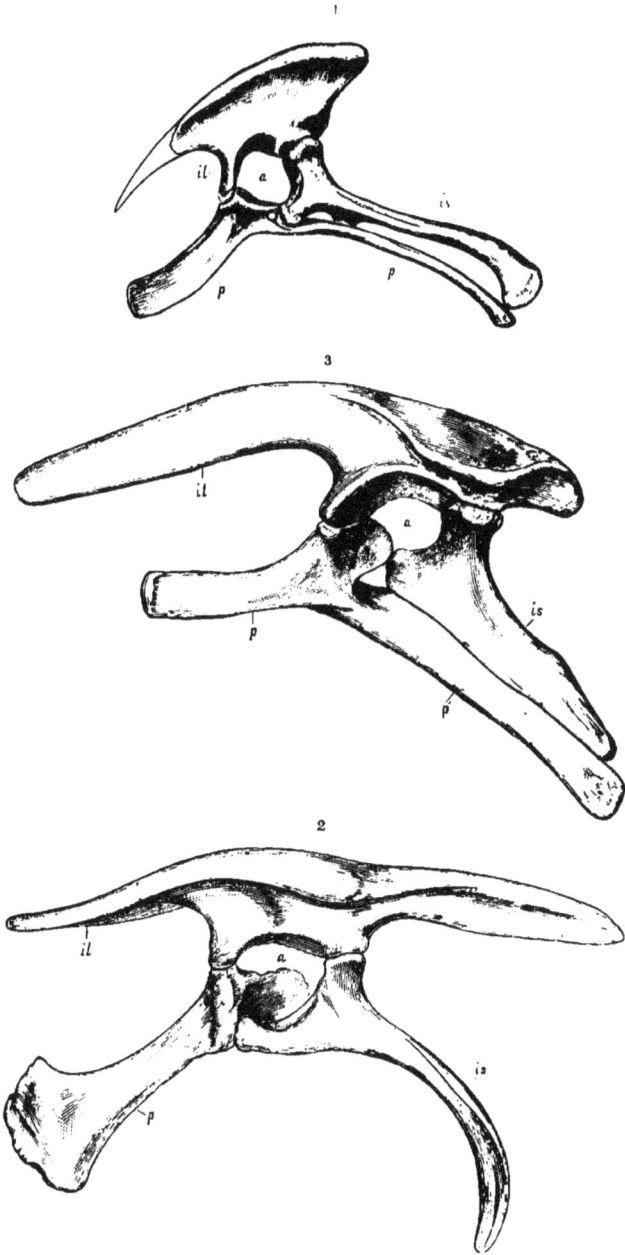

PELVES OF PREDENTATE DINOSAURS. CAMPTOSAURUS, STEGOSAURUS AND STERRHOLOPHUS

PLATE LXXXII.

PLATE LXXXII.

European Dinosaurs.—Theropoda.

Compsognathidæ.

   408

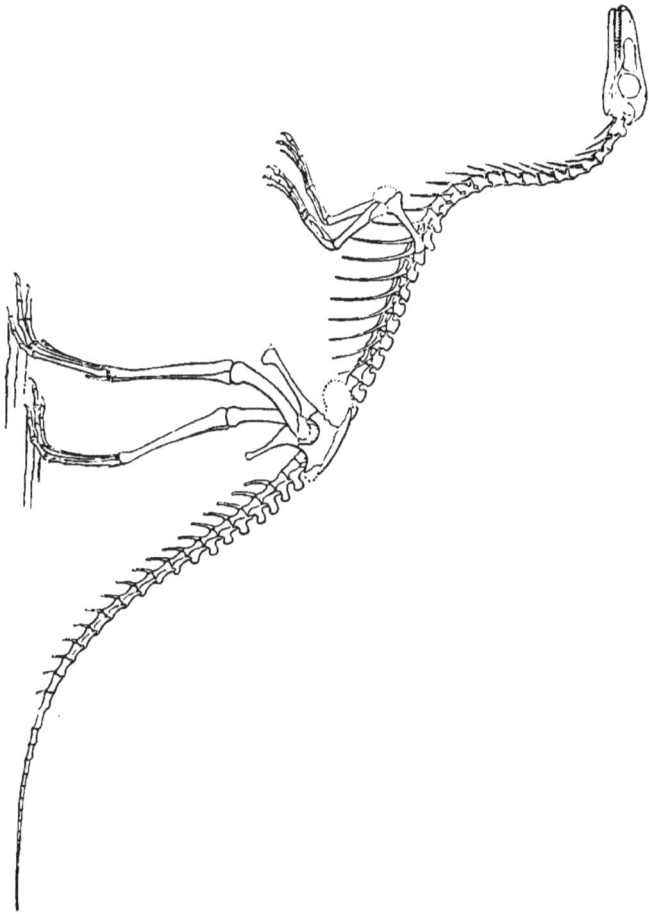

PLATE LXXXIII.

PLATE LXXXIII.

European Dinosaurs.—Predentata.

Scelidosauridæ.

One-eighteenth natural size.
Jurassic, England.

410

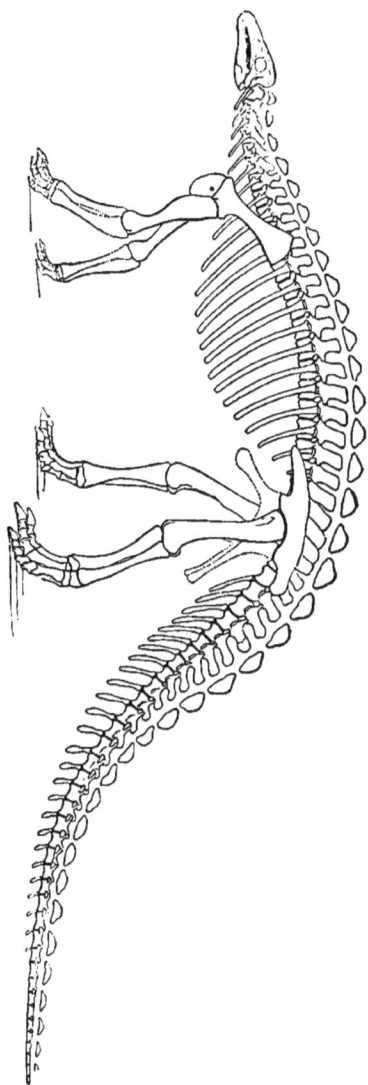

PLATE LXXXIV.

411

PLATE LXXXIV.

EUROPEAN DINOSAURS.—PREDENTATA.

HYPSILOPHODONTIDÆ.

412

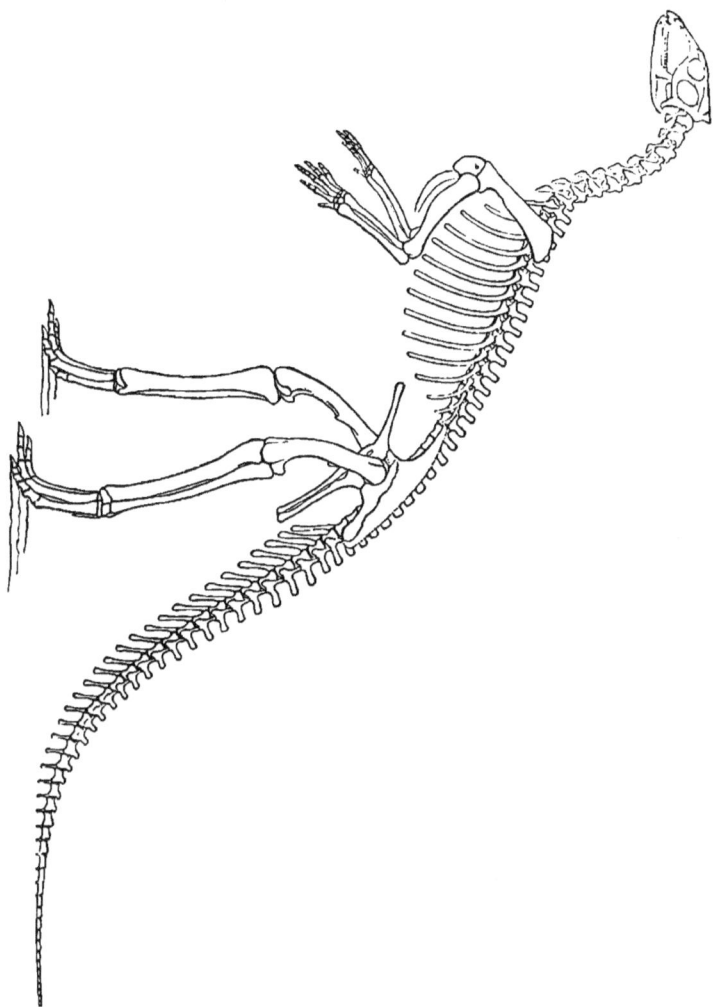

PLATE LXXXV.

PLATE LXXXV.

European Dinosaurs.—Predentata.

Iguanodontidæ.

414

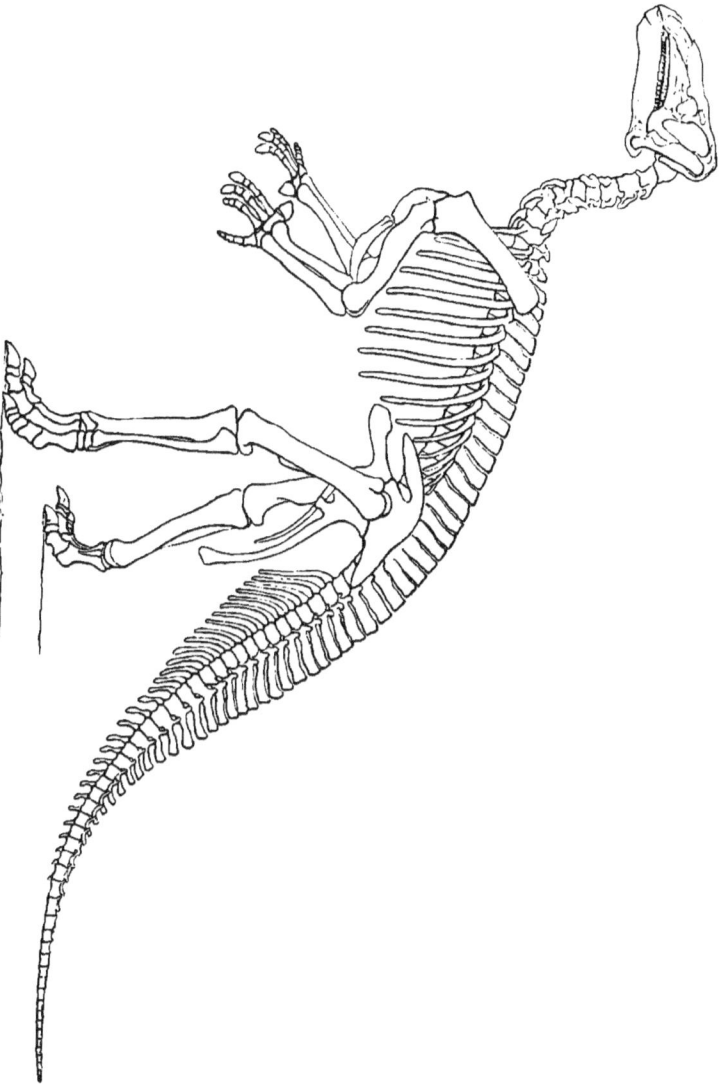

RESTORATION OF IGUANODON BERNISSARTENSIS Boulenger.

One-fortieth natural size. Wealden, Belgium.